STRUCTURAL BEHAVIOUR OF CONCRETE WITH COARSE LIGHTWEIGHT AGGREGATES

A.A.BALKEMA / ROTTERDAM / BROOKFIELD / 1995

CUR Report 173

Centre for Civil Engineering Research and Codes

Published and distributed for CUR, Gouda by
A.A. Balkema, P.O. Box 1675, 3000 BR Rotterdam, Netherlands (Fax: +31.10.4135947)
A.A. Balkema Publishers, Old Post Road, Brookfield, VT 05036, USA (Fax: 802.276.3837)

ISBN 90 5410 625 5

PREFACE

CUR Research Committee C 75 "Lightweight Concrete" was established in 1989. Its task was to conduct research into the material properties that are important for the structural behaviour of lightweight concrete, to draw up a CUR Recommendation with rules for design, calculation and construction of concrete structures of lightweight concrete, and to summarize the research results in a CUR report.

The need for further research and for specific technical principles for the application of lightweight concrete arose from the availability of new lightweight aggregates, such as Lytag and Aardelite. Experiences with concrete containing these lightweight aggregates led to the conclusion that a number of properties deviate from the types of lightweight concrete applied until then. These experiences have been included in CUR report 89-3 "Lytag as an aggregate for concrete" by CUR preadvisory committee PD 9 "Construction with concrete with lightweight aggregates". This report formed the basis for the investigations carried out under supervision of CUR Research Committee C 75.

CUR Research Committee C 75 also paid considerable attention to the durability aspects of lightweight concrete. Working group 1 "Durability of lightweight concrete" thoroughly investigated these aspects, which resulted in an account that has been included in this final report.

The results of the research led to CUR Recommendation 39 "Concrete with coarse lightweight aggregates", which includes requirements and calculation rules supplementary to NEN 6720 "Technical Principles for Building Structures TGB 1990-Regulations for concrete. Structural requirements and calculation methods (VBC 1990)", NEN 6722 "Regulations for concrete. Execution (VBU 1988)" and NEN 5950 "Regulations for concrete. Technology. Requirements, manufacturing and testing (VBT 1986)". The CUR Recommendation was drawn up by working group 2 "Regulations" and 3 "Technology and Execution", respectively.

At the time of publication of this integral final report, which also includes the background of CUR Recommendation 39, CUR Research Committee C 75 and the working groups are composed as follows:

ir. J.H.J. MANHOUDT, Chairman
ir. P. DE JONG, Secretary
ir. D.W. BILDERBEEK
ir. P. EGGERMONT
dr.ir. J.W. FRÉNAY
ir. M. LEEWIS
ing. B.J.H.M. VAN DER SANDE-SCHREURS
ing. R.J. UITERWIJK
ing. H. VOORTMAN
ir. F.B.J. GIJSBERS, Rapporteur
ir. J.A. DEN UIJL, Rapporteur
prof.dr.ir. J.C. WALRAVEN, Rapporteur

ing. A. Gerritse, Corresponding member
ing. A.C. Fuchs, Coordinator
prof.ir. J.H. van Loenen, Mentor

Working group 1 "Durability of lightweight concrete"
dr. R.F.M. Bakker, Chairman
dr.ir. H.A.W. Cornelissen, Secretary
ir. P. Eggermont
ing. R.J. Uiterwijk
C.J.J. Castenmiller, Rapporteur
ir. A.J.M. Siemes, Rapporteur

Working group 2 "Regulations"
ir. G.L.H.M. Henkens, Chairman
ir. P. de Jong, Secretary and Rapporteur
ir. P. Eggermont
ir. F.B.J. Gijsbers
ir. J. Stroband
ing. A.R. Kerp, Coordinator

Working group 3 "Technology and execution"
ing. R.J. Uiterwijk, Chairman
ir. C.A. van der Steen, Secretary and Rapporteur
ing. W.F. Lolkus
ing. B.J.H.M. van der Sande-Schreurs
T.W.M. Schuiling
ing. H. Voortman

The research was carried out by the Stevin Laboratory of the Delft University of Technology, and TNO Building and Construction Research. Ir. J. Stroband and ir. J.A. den Uijl of the TU Delft and ir. N.M. Naaktgeboren of TNO Building and Construction Research have largely contributed to the success of this investigation. The literature study carried out for chapter 6 "Durability" was supervised and partly written by prof. dr. ir. J.C. Walraven. Ir. F.B.J. Gijsbers edited the report.
The publication of this report and CUR Recommendation 39 meets with the specific need for further knowledge and technical principles on the basis of which a safe and accepted application of lightweight concrete will be possible.

The CUR is indebted to NOVEM, Ministry of Economic Affairs, Provincial Department of Public Works Gelderland, Civil Engineering Division of the Ministry of Transport, Public Works and Water Management, Vliegasunie B.V., and VASIM B.V. for their financial contributions that helped to make this research possible.

June 1995   The Executive Committee of CUR

CONTENTS

5

SUMMARY

STRUCTURAL BEHAVIOUR OF CONCRETE WITH COARSE LIGHTWEIGHT AGGREGATES

This report contains a summary of the various internal reports on the research executed by CUR Research Committee C 75 "Lightweight concrete". The aim of this research was to establish calculation rules, to be laid down in a CUR-Recommendation for concrete structures with coarse lightweight aggregates. To obtain these rules a number of structural aspects was investigated both experimentally and numerically. Both the research and the CUR-Recommendation were limited to concrete with Lytag, Aardelite and Liapor as lightweight aggregates. The investigated structural aspects concerned the splitting, the bond and the shear behaviour. The various sub-investigations are discussed separately in this report. Also the durability aspect of lightweight concrete is taken into account. This aspect has been investigated on the basis of a literature study.

The behaviour of lightweight concrete deviates from gravel concrete in a number of points. These differences are related to the difference in mass volume of both materials. This agrees with the system used in the European standard for lightweight concrete structures, which is being developed, so a future switch to the European standard is anticipated. The research results have been included in CUR-Recommendation 39 "Concrete with coarse lightweight aggregates". This CUR-Recommendation contains supplementary regulations to NEN 6720 "Regulations for concrete – Structural requirements and calculation methods (VBC 1990)". Apart from the structural rules, also supplementary regulations with regard to technology and construction have been included in this CUR-Recommendation.

LIST OF SYMBOLS

A	area of the loading plate
A_b	cross-sectional area of the test specimen
A_0	cross-sectional area of the test specimen as far as it is concentric with the loading plate
b	width of the test specimen
b_w	web width of the I-shaped beam
c	concrete cover
d	effective height of the cross section of the beam, thickness of the test specimen
E'_b	modulus of elasticity of concrete
E_s	modulus of elasticity of reinforcing steel
e	eccentricity
F_u	failure load
$F_{u\,ber}$	calculated failure load
$F_{u\,exp}$	experimental failure load
f_b	uniaxial tensile strength, (splitting) tensile strength
f'_b	uniaxial compressive strength, prism compressive strength
f'_{b0}	ultimate stress under the loading plate
f'_{ck}	cube compressive strength
G_f	crack energy
h	height of the test specimen
k	width of the loading plate
l	length, length of the loading plate
l_1	lap length
M_u	failure moment
$M_{u\,ber}$	calculated failure moment
$M_{u\,exp}$	experimental failure moment
N_r	force at which cracking occurs
N_s	force in the reinforcing steel
V_u	shear force in the failure stage
w	crack width
μ	coefficient of friction
ν	Poisson's ratio
ρ	density of concrete
σ'_b	concrete stress
σ'_s	steel stress
τ_{exp}	maximum experimental shear stress
τ_s	maximum shear stress by shear reinforcements
τ_u	bond stress, bond strength
τ_{VBC}	shear strength according to VBC 1990
τ_1	shear strength without shear reinforcement
τ_2	shear strength with shear reinforcement
ω_0	longitudinal reinforcement percentage
Δl	elongation

CHAPTER 1

INTRODUCTION

In 1987 CUR established pre-advisory committee PD 9 "Construction with concrete with lightweight aggregates". The task of this committee was to survey and evaluate the experiences with lightweight concrete in the Netherlands. The committee was also expected to advise on the desired, CUR-related, research and regulation activities concerning the application of concrete with lightweight aggregates. On the basis of the advice of the committee, which was reported in [1], the CUR established Committee C 75 "Lightweight Concrete".

The present CUR report contains a brief summary of the research carried out by this committee. This research included:

- research into the properties of concrete with coarse lightweight aggregates which are of importance with regard to its structural behaviour;
- drafting a CUR-Recommendation including design and calculation rules for structures of concrete with coarse lightweight aggregates.

The CUR-Recommendation to be formulated was to be in keeping with the Dutch code NEN 6720 (published in 1991) "Technical Principles for Building Structures TGB 1990 – Regulations for Concrete. Structural requirements and calculation methods (VBC 1990)". The research and the CUR-Recommendation have been limited to concrete with the coarse lightweight aggregates Lytag, Aardelite and Liapor. The product specifications of these aggregates are given in appendix A. For purposes of comparison, experiments on gravel concrete have also been carried out. The application of coarse lightweight aggregates also has consequences for the technology and the execution. These consequences have been translated into additions to the VBT 1986 and the VBU 1988, which have also been included in the CUR-Recommendation. The technological and executional aspects are, however, not discussed in this report.

The structural properties have been investigated both experimentally and numerically. In the experimental research a number of material properties have been established for the three types of lightweight concrete investigated. A survey of these properties is given in chapter 2. The experimental research concerned two strength classes, with an average target short-term cube strength of 30 N/mm^2 and 60 N/mm^2, respectively. Although the actual strengths slightly deviate from these target strengths, the test results are regarded to be representative for the strength classes B 30 and B 60. These designations have therefore been used throughout the report.

In the numerical research the finite element method DIANA has been used. The numerical research was aimed at the possibility to gain insight in a numerical way into the structural behaviour of concrete with other aggregates, when the basic characteristics are known from experimental research. Therefore the extent to which the numerically determined structural behaviour is influenced by variations in the basic characteristics of concrete has been verified throughout the numerical research.

The splitting behaviour, bond behaviour and shear behaviour of concrete is discussed in the chapters 3, 4 and 5 respectively. Also the durability aspect of lightweight concrete was included in the research. This aspect was considered on the basis of a literature study and is discussed in chapter 6. Chapter 7 concerns the structural regulations for structures made of concrete with the coarse lightweight aggregates Lytag, Aardelite or Liapor, that resulted from the investigations.

This summarizing report is based on the internal reports of the committee, to which the reader is referred throughout the text for more detailed information. These internal reports are listed in appendix B.

CHAPTER 2

MATERIAL PROPERTIES

Part of the experimental research was devoted to determining various mechanical properties of the types of concrete investigated, namely:

- cube compressive strength f'_{ck}
- prism compressive strength f'_b
- (splitting) tensile strength f_b
- modulus of elasticity E'_b
- Poisson's ratio v
- crack energy G_f

For concrete with hard, dense aggregates, such as gravel, the VCB 1990 gives relations by means of which the uniaxial tensile strength (prism compressive strength), the (splitting) tensile strength and the modulus of elasticity can be derived from the cube compressive strength. These relations are:

$$f'_b = 0.85\,f'_{ck}$$

$$f_b = 1.05 + 0.05\,f'_{ck} \quad \text{with } f_b \text{ and } 5\,f'_{ck} \text{ in N/mm}^2$$

$$E'_b = 22250 + 250\,f'_{ck} \quad \text{with } E'_b \text{ and } 5\,f'_{ck} \text{ in N/mm}^2$$

Figures 1 to 4 show the experimental results together with these relations for concrete with the aggregates Lytag, Aardelite, Liapor and gravel, respectively. The values indicated in these figures are the average values of three or six test specimen. With the exception of the modulus of elasticity, the differences between gravel concrete and the investigated types of lightweight concrete are small.

The splitting tensile strength measured from cubes, which were cured in dry circumstances, turned out to be 10% lower than the splitting tensile strength measured from cubes, which were cured under water. This has to be attributed to the internal stresses that arise from drying of the test specimen. Should the irregular drying process be continued a further reduction in the tensile strength might occur due to increasing internal damage. For this reason the tensile strength of concrete with lightweight aggregates is usually set slightly lower than that of gravel concrete.

The deviations of the design values of the material properties of lightweight concrete as compared to those of gravel concrete, will be discussed in chapter 7.

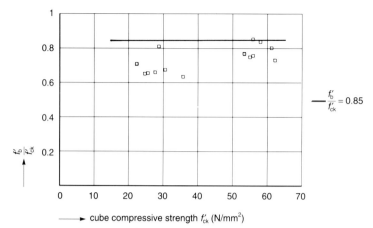

a. prism compressive strength as a function of the cube compressive strength

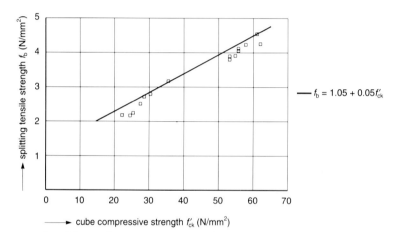

b. splitting tensile strength as a function of the cube compressive strength

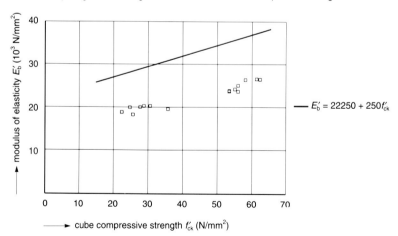

c. modulus of elasticity as a function of the cube compressive strength

Fig. 1.   Material properties of Lytag concrete.

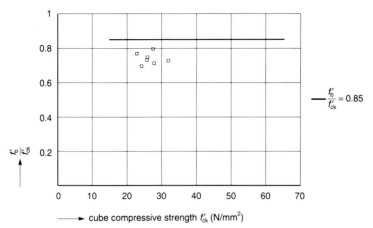

a. prism compressive strength as a function of the cube compressive strength

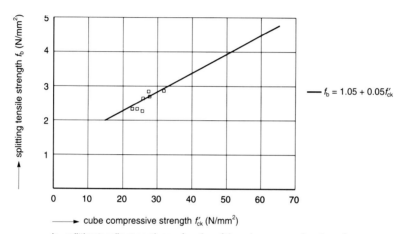

b. splitting tensile strength as a function of the cube compressive strength

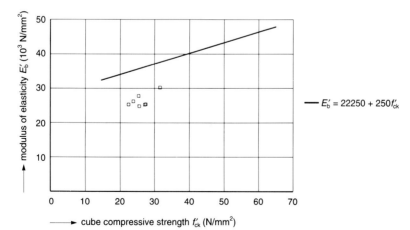

c. modulus of elasticity as a function of the cube compressive strength

Fig. 2.   Material properties of Aardelite concrete.

15

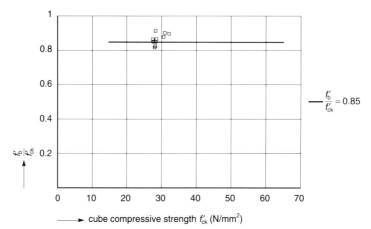

a. prism compressive strength as a function of the cube compressive strength

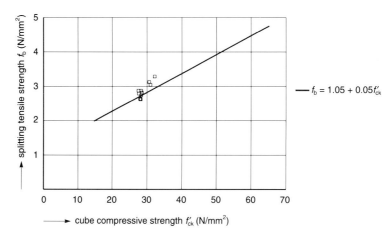

b. splitting tensile strength as a function of the cube compressive strength

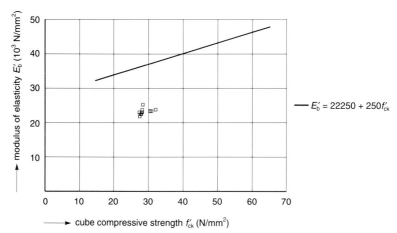

c. modulus of elasticity as a function of the cube compressive strength

Fig. 3.   Material properties of Liapor concrete.

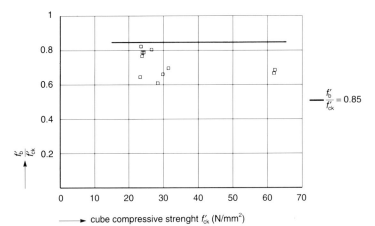

a. prism compressive strength as a function of the cube compressive strength

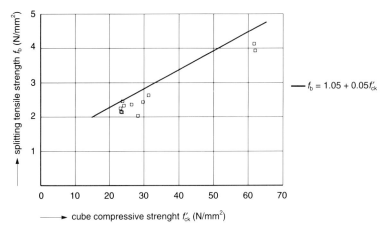

b. splitting tensile strength as a function of the cube compressive strength

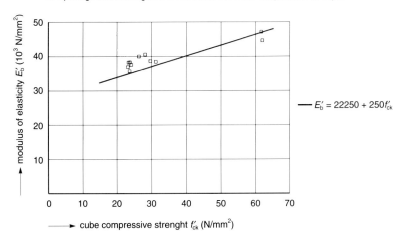

c. modulus of elasticity as a function of the cube compressive strength

Fig. 4. Material properties of gravel concrete.

17

CHAPTER 3

SPLITTING BEHAVIOUR

3.1   Splitting behaviour of plain concrete panels

3.1.1 *General*

The research into the splitting behaviour of plain concrete panels concerned the behaviour of the unreinforced material during a bi-axial tension-compression combination, making use of a combined numerical and experimental analysis. The aim of this research was to gain insight into the splitting behaviour of light-weight concrete and the way in which this splitting behaviour is influenced by the fact that cracks in lightweight concrete usually run through the aggregate.

3.1.2 *Experimental research into the splitting behaviour*

The experimental research into the splitting behaviour [2] has been carried out on plain concrete panels, which were in-plane loaded with a concentrated load. The load was applied in a deflection controlled way in order to make it possible to continue to monitor the behaviour beyond the maximum load. The experimental research included eight series of five test specimen. The influence of the follow-ing parameters was investigated:

- the dimensions of the test specimen (the width-height relation of the test speci-men is 1 : 2);
- the dimensions of the loading plate;
- the type of aggregate (Lytag, Aardelite, Liapor, gravel);
- the strength class of the concrete (B 30, B 60);
- the loading rate.

Table 1 shows a survey of the test results. Figure 5 shows the test set-up. Between the loading plate and the test specimen cardboard was applied as a load distributing layer, after preliminary experimental research showed that this hardly prevented the transverse displacements. During the tests the following measurements were carried out at the front and the back side of the test specimen (see also figure 5):

- the magnitude of the load;
- the transverse deformation over the height in the middle of the test specimen;
- the longitudinal deformation halfway the height;
- the total longitudinal deformation between the loading plates.

The positions of the measuring instruments are based on the expected behaviour of the test specimen which was umedically determined [3].
In the following the strength, deformation and failure behaviour of the test speci-men have been discussed.

Table 1. Summary of test results of the panels with concentrated load.

nr.	width b (mm)	width of loading plate k (mm)	k/b	strain rate**	failure load F_u (kN)	failure stress f'_{b0} (N/mm²)	cube compressive strength f'_{ck} (N/mm²)	f'_{b0}/f'_{ck}
gravel concrete B 30							31.2	
11	500	150	0.3	100	539	24.0		0.77
12	300	150	0.5	100	428	19.0		0.61
13	300	90	0.3	100	313	23.2		0.74
14	300	30	0.1	100	215	47.8		1.53
15	150	45	0.3	100	182	27.0		0.86
Lytag concrete B 60							55.6	
21	500	150	0.3	100	963	42.8		0.77
22	300	150	0.5	100	845	37.6		0.68
23	300	90	0.3	100	602	44.6		0.80
24	300	30	0.1	100	382	84.9		1.53
25	150	45	0.3	100	286	42.4		0.76
Lytag concrete B 30							28.6	
31	500	150	0.3	100	514	22.8		0.80
32	300	150	0.5	100	474	21.1		0.74
33	300	90	0.3	100	313	23.2		0.81
34	300	30	0.1	100	169	37.6		1.31
35	150	45	0.3	100	185	27.4		0.96
gravel concrete B 60							61.5	
41	500	150	0.3	100	814	36.2		0.59
42	300	150	0.5	100	697	31.0		0.50
43	300	90	0.3	100	522	38.7		0.63
44	300	30	0.1	100	336	74.7		1.21
45	150	45	0.3	100	286	42.4		0.69
Aardelite concrete B 30							31.8	
51	500	150	0.3	100	501	22.3		0.70
52	300	150	0.5	100	463	20.6		0.65
53	300	90	0.3	100	332	24.6		0.77
54	300	30	0.1	100	193	42.9		1.35
55	150	45	0.3	100	189	28.0		0.88
Lytag concrete B 60							57.7	
61	500	150	0.3	100	949	42.2		0.73
62	300	90	0.3	1	507	37.6		0.65
63	300	90	0.3	100	629	46.6		0.81
64	300	90	0.3	10000	700	51.9		0.90
65	150	45	0.3	100	290	43.0		0.74
Lytag concrete B 60							60.9	
71	500	150	0.3	100	983	43.7		0.72
72	300	90	0.3	1	551	40.8		0.67
73	300	90	0.3	100	602	44.6		0.73
74	300	90	0.3	10000	658	48.7		0.80
75	150	45	0.3	100	287	42.5		0.70
Liapor concrete B 30							28.5	
81	500	150	0.3	100	562	25.0		0.88
82	300	150	0.5	100	475	21.1		0.74
83	300	90	0.3	100	327	24.2		0.85
84	300	30	0.1	100	190	42.2		1.48
85	150	45	0.3	100	176	26.1		0.91

* dimensions test specimen $b \times h \times d$, in which: $h = 2d$ en $d = 150$ mm

** deformation rate in 10^{-6} mm/s at the inductive recorders

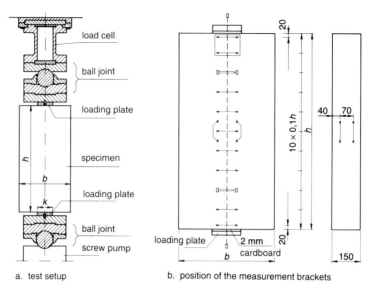

a. test setup   b. position of the measurement brackets

Fig. 5.   Experimental research into the plates loaded for splitting.

Strength

The registered failure loads are summarized in table 1.

The most important factor influencing the results is the ratio of the area of the loading plate to the cross-sectional area of the test specimen. Figure 6 shows the relation between the ultimate stress under the loading plate and the cube compressive strength of the concrete as a function of the relation of the cross-sectional area of the test specimen and the area of the loading plate. This relation is valid for the test specimen with the dimensions $b \times h = 300$ mm $\times 600$ mm. In the investigated range the relation is practically linear.

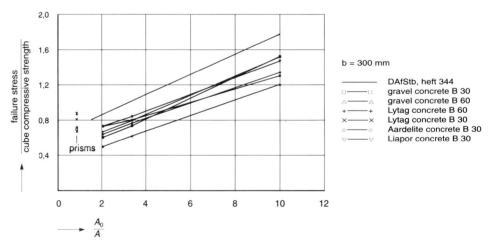

Fig. 6.   Influence of the width of the loading plate as a function of the cube compressive strength.

20

With a ratio of $A_b/A = 2$, i.e. with a loading plate half as wide as the test specimen, the ultimate stress is lower than the prism compressive strength of the concrete. The smaller the width of the loading plate, the higher the ultimate stress.
The test results confirm the known strong influence of the width of the loading
plate on the ultimate stress. Figure 6 does not show a clear difference between
gravel concrete and lightweight concrete. The ratio given by WURM and DASCHNER
renders a considerably higher failure load [4]. This difference may be ascribed to
the fact that the test specimen in this particular investigation were loaded at one
side with a concentrated load. Moreover glued loading plates were used, which
prevented the transverse deformation of the concrete under the loading plate. With
regard to the influence of the type of concrete, the conclusion can be drawn that the
behaviour of lightweight concrete is not less favourable than that of gravel concrete. The influence of the strength class is not entirely unambiguous. In most cases
the relative ultimate stress with strength class B 60 is lower than with strength class
B 30, which is explicable, given the relation between the splitting tensile strength
and the cube compressive strength. However, with Lytag concrete, a narrow loading plate rendered the opposite result. This may be due to the scatter in the observations, but the number of test specimen is too small to exclude other causes.
With respect to the influence of the dimensions of the test specimen, there
appears to be a certain scale effect, which in general leads to a relatively higher
strength when the dimensions of the test specimen are smaller. Lytag concrete B
60, however, does not show this effect. This, too, may be attributed to the scatter
in the observations, but other causes cannot be excluded.
As regards the influence of the loading rate, the strength appears to decrease more
or less linearly with the logarithm of the load duration. With a load duration of 17
hours the resulting strength was 30 % lower than with a load duration of 6 seconds.
Whether this trend will continue cannot be concluded from this investigation.

Deformation

The transverse deformation in the vertical section below the centre of the load
presents a characteristic picture. In figure 7 this is shown for one of the test specimen in a three-dimensional diagram. The load steps are plotted on the horizontal
axis. For every load step the measured elongation over the measuring length is plotted vertically for the various reference points. The third axis gives the distance
between the reference points and the top of the test specimen. At some points on the
horizontal axis the applied load has been indicated. The load increases to the maximum value Fu (in this case 313 kN) and subsequently decreases again. The figure
shows that the deformations at first occur only in the vicinity of the load at the top
and bottom of the test specimen. The deformation halfway the height remains far
behind. This continues until beyond the maximum load, up to the point where the
deformation over the entire height increases rather suddenly. In general, the tensile
strength is reached at an early stage in the vicinity of the applied load, after which
the behaviour is determined locally by tension-softening. Halfway the height the
strain falls behind, but at maximum load also the tensile strength is reached.

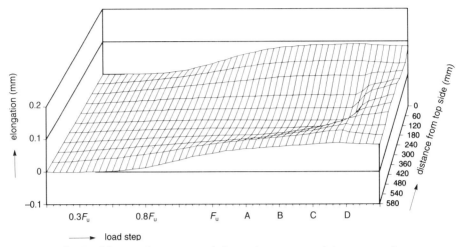

Fig. 7. Measured transverse deformations at one of the test specimen.

Failure behaviour

The concentrated load generates high compressive stresses under the loading plate and splitting stresses in the middle of the test specimen. Thus there are two possible failure mechanisms: crushing of the concrete under the loading plate and splitting of the test specimen. With the investigated test specimen only the second failure mechanism occurred. When splitting reinforcement is applied this failure mechanism can be suppressed, in which case the first failure mechanism will occur. Figure 8 shows a survey of the observed cracking. In most cases one continuing crack occurred, whereas in some cases cracks also occurred next to the loading plate. Because the deformation was controlled, the failure occurred in the descending part of the load-deformation diagram.

3.1.3 *Numerical research into the splitting behaviour*

A preliminary investigation [3] was carried out to examine the way in which the elements would behave. Moreover, this preliminary study formed the basis for the selection of the element grid to be used in the actual analysis and the modelling of the material behaviour. The numerical results were evaluated on the basis of data collected from literature on gravel concrete [4]. The results of the preliminary investigation partly formed the basis for the measuring programme used in the experimental research.

Six of the experiments were numerically simulated [5], which means that a representative selection of the experimental programme was made. The simulated tests are included in table 2. In view of symmetry, a quarter of the entire panel has been analyzed. The loading plate was assumed to be infinitely rigid and rested on the test specimen without any friction.

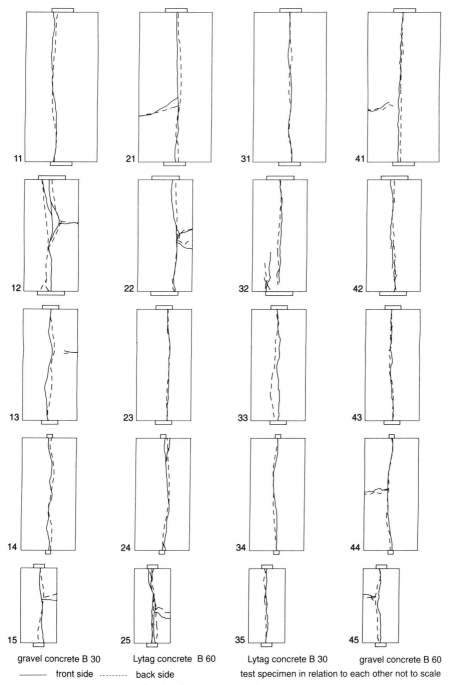

11	21	31	41
12	22	32	42
13	23	33	43
14	24	34	44
15	25	35	45

| gravel concrete B 30 | Lytag concrete B 60 | Lytag concrete B 30 | gravel concrete B 60 |

——— front side ········· back side   test specimen in relation to each other not to scale

Fig. 8. Observed crack patterns.

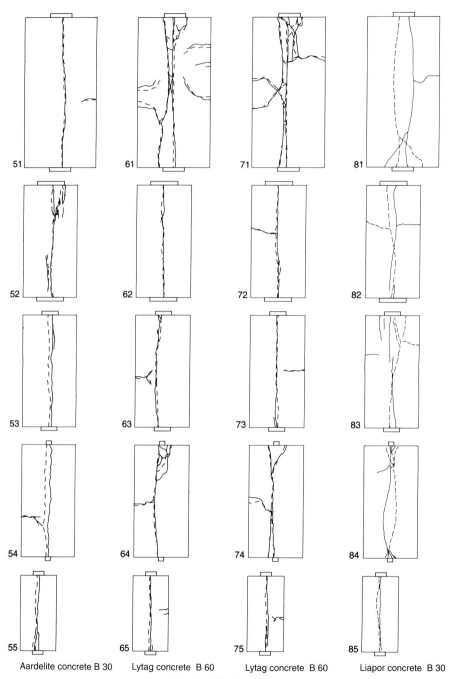

Fig. 8. Continuation.

Table 2. Summary of the numeric results of the panels loaded for splitting.

test	aggregate	strength class	specimen $b \times h$ (mm)	loading plate k/b	experimental failure load $F_{u\,exp}$ (kN)	calculated failure load $F_{u\,ber}$ (kN)	$F_{u\,ber}/F_{u\,exp}$
13	gravel	B 30	300×600	0.3	313	394	1.26
23	Lytag	B 60	300×600	0.3	602	438	0.73
31	Lytag	B 30	500×1000	0.3	514	547	1.06
32	Lytag	B 30	300×600	0.5	474	454	0.96
33	Lytag	B 30	300×600	0.3	313	315	1.01
34	Lytag	B 30	300×600	0.1	169	209	1.24

The input parameters are in accordance with the experimental research, namely:

- the uniaxial tensile strength f_b, determined from the cube strength f'_{ck} according to:
  for gravel concrete: $\quad f_b = 1 + 0.05\,f'_{ck}$;
  for lightweight concrete: $\quad f_b = 0.9(1 + 0.05\,f'_{ck})$;
- the uniaxial compressive strength f'_b, determined from the cube compressive strength f'_{ck} according to:
  $f'_b = 0.8\,f'_{ck}$;
- the modulus of elasticity E'_b, determined from the experimentally determined stress-strain relations for compressed prisms;
- Poisson's ratio, which was set at 0.2, with the exception of Lytag B 60, for which 0.25 was chosen on the basis of the test results;
- the crack model, using the tension-softening model according to CORNELISSEN, HORDIJK and REINHARDT [6];
- the fracture model, using a linear relation between the tensile strength and the compressive stress in the perpendicular direction.

The different input parameters have partly been chosen on the basis of conducted parameter analyses, in which the influence of the modulus of elasticity, Poisson's ratio and the boundary conditions at the load introduction were investigated.

The results of the numerical simulation of one of the analyzed test specimen are shown in the figures 9 to 11. These figures also show the experimentally determined results. In figure 9 the transverse displacements over the height of the test specimen are given at approximately 30 % and 60 % of the experimental failure load, and in figure 10 at 100 % of the numerical failure load. At lower load levels the results correspond quite well. At higher load levels fairly large differences occur immediately under the loading plate. This may partly be due to the different boundary conditions (in the test cardboard was used, in the calculation a frictionless connection between the loading plate and the test specimen was assumed). The numerical analysis of a test specimen in which transverse deformations at the loading plate were completely prevented, showed a somewhat better correspondence.

Figure 11 shows the calculated transverse stress over the height of the test specimen for three load levels. This figure clearly shows that when the failure load is reached (316 kN), the tensile stresses over part of the height are situated at the beginning of the softening branch. This is because the stresses belonging to the failure load in the area between 80 mm and 200 mm below the loading plate are lower than in the preceding load step.

In table 2 the experimentally and numerically determined failure loads are compared. On average the correspondence appears to be quite well, though occasionally considerable differences occur. This may be due to a scatter in the test results, but a difference in behaviour of the material under tension may also be the cause. Also the differences in the boundary conditions at the load introduction may have played a role.

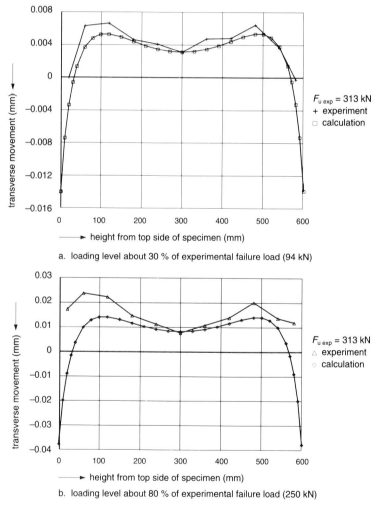

a. loading level about 30 % of experimental failure load (94 kN)

b. loading level about 80 % of experimental failure load (250 kN)

Fig. 9.   Calculated and measured transverse displacements over the height of the test specimen.

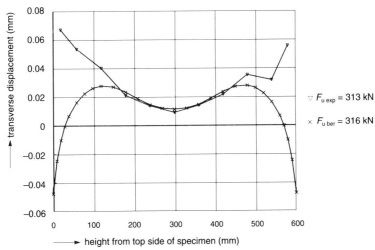

Fig. 10. Calculated and measured transverse displacements over the height of the test specimen at failure.

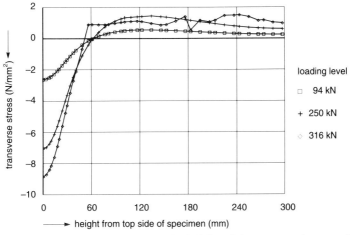

Fig. 11. Calculated transverse stresses over the height of the test specimen at three load levels.

Figure 12 shows the calculated and the measured influence of the width of the loading plate on the ultimate stress. The figure also includes this relation according to WURM and DASCHNER.

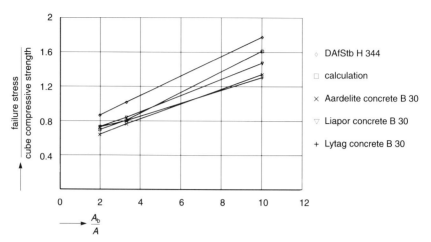

Fig. 12. Influence of the loading plate width on the ultimate stress.

Table 3. Numerically examined combinations of parameters.

calculation	tensile strength (N/mm²)	crack energy (J/m²)	calculated failure load (kN)
reference	2.2	60	315
1	1.4	60	254
2	3.0	60	348
3	2.2	100	336
4	3.0	100	373
5	2.2	20	256
6	1.4	20	206

Finally, a numerical parameter analysis has been carried out [7] to examine the influence of the tensile strength and the crack energy. Table 3 shows the investigated combinations. The calculated failure loads for these combinations are shown in figure 13. A decrease in crack energy results, as expected, in a lower strength and a more brittle behaviour, in particular for values lower than the experimentally investigated range. The parameter analysis warrants the conclusion that in order to predict the bi-axial behaviour, both the tensile strength and the tension-softening behaviour need to be known because of their large influence on the strength and the failure behaviour.

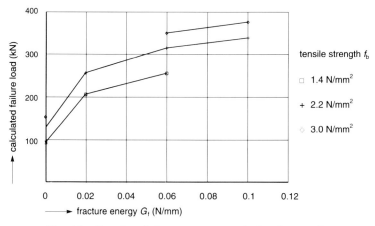

Fig. 13. Results of the parameter study for the panels.

3.1.4 *Conclusions*

The most important conclusions to be drawn from this part of the investigation are:

- No significant differences have emerged between gravel concrete and the types of lightweight concrete investigated. This is valid for the strength as well as for the deformation behaviour.
- The density of the applied lightweight aggregates varied from about 900 to about 1600 kg/m³. This covers the range of structural lightweight concrete.
- The strength is mainly influenced by the relation of the loading plate width to the total width of the test specimen. In addition, the strength is influenced by the loading rate and, to a small extent, by the dimensions of the test specimen. The latter implies that scale effects do play a role.
- Tension-softening plays an important role. Long before the failure stage is reached, softening occurs, accompanied by a redistribution of the stresses. This could be verified with the help of numerical research.
- The conditions at the load introduction in the tests and in the calculations are not fully comparable (cardboard as intermediate layer and completely friction-less, respectively). Partly for this reason differences occur between the numerical and the experimental results, particularly in the area immediately under the load introduction.
- It appears to be very well possible to numerically simulate the experimentally acquired results for the investigated panels. Therefore, in this case, the numerical analysis is a suitable tool to establish the effect of variations in the material properties. When the material behaviour under tension is known, tension-softening included, reliable numerical predictions can be made about the splitting behaviour, also for other types of lightweight concrete than the ones investigated here.

3.2   Splitting behaviour of eccentrically loaded plain concrete prisms

3.2.1 *General*

Although the crack energy and the modulus of elasticity for the types of light-weight concrete investigated in the previously described research into panels are considerably lower than for gravel concrete, no important differences in the failure behaviour of these panels has been established. Apparently these material properties did not play a dominant role in these tests. To verify whether this is also valid if the load can be distributed in two directions, prisms were tested under concentrated loading conditions (see figure 14). This part of the investigation did not include any numerical research.

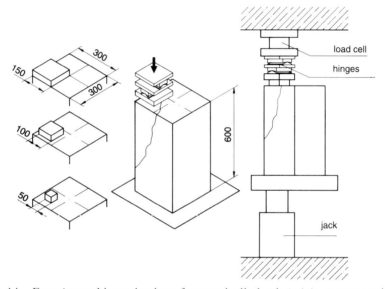

Fig. 14.   Experimental investigation of eccentrically loaded plain concrete prisms.

3.2.2 *Experimental research*

The experimental research [8] included five series of three test specimen. The influence of the following parameters was investigated:

- the dimensions of the loading plate;
- the aggregate (Lytag, Aardelite, Liapor, gravel);
- the strength class of the concrete (only for Lytag).

The results of the investigation are shown in table 4. This table gives the failure load, the ultimate stress f'_{b0} under the loading plate and the relation of the ultimate stress to the cube and prisms compressive strengths respectively. The prisms collapsed either because of shear of a corner or because of splitting (see figure 15).

30

Table 4.   Summary of the test results of the prisms.

test	loading plate $k \times l$ (mm)	cube compressive strength f'_{ck} (N/mm²)	prism compressive strength f'_{b} (N/mm²)	failure load F_{u} (kN)	ultimate stress f'_{b0} (N/mm²)	f'_{b0}/f'_{ck}	f'_{b0}/f'_{b}
Lytag concrete B 60		53.1	41.1				
501	150 × 150			977	43.4	0.8	1.1
502	100 × 100			633	63.3	1.2	1.5
503	50 × 50			272	108.8	2.0	2.7
Liapor concrete B 30		28.0	23.0				
601	150 × 150			605	26.9	1.0	1.2
602	100 × 100			346	34.6	1.2	1.5
603	50 × 50			132	52.8	1.9	2.3
Lytag concrete B 30		22.2	15.9				
701	150 × 150			590	26.2	1.2	1.7
702	100 × 100			342	34.2	1.5	2.2
703	50 × 50			150	60.0	2.7	3.8
Aardelite concrete B 30		23.1	17.5				
901	150 × 150			630	28.0	1.2	1.6
902	100 × 100			370	37.0	1.6	2.1
903	50 × 50			167	66.8	2.9	3.8
gravel concrete B 30		23.5	19.3				
1101	150 × 150			630	28.0	1.2	1.5
1102	100 × 100			416	41.6	1.8	2.2
1103	50 × 50			210	84.0	3.6	4.4

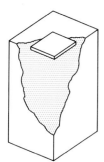
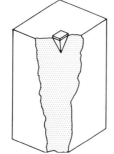

Fig. 15.   Some characteristic failure modes.

Figure 16 shows the relation failure load/cube compressive strength as a function of the relation A/A_0. The influence of the dimensions of the loading plate correspond to the known trend, in the sense that the ultimate stress increases less than proportionally with the relation A/A_0. In this A is the area of the loading plate and A_0 the cross-sectional area of the test specimen as far as it is concentric with the loading plate.

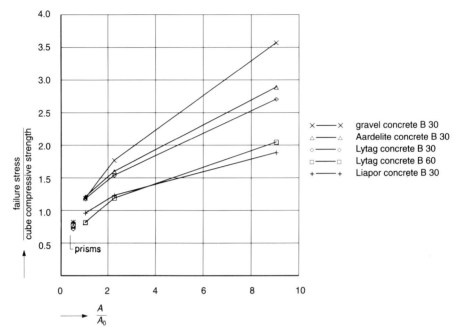

Fig. 16. Failure stress as a function of the cube compressive strength.

However, the ultimate stress of lightweight concrete also appears to increase more slowly than that of gravel concrete. It has been established that the ultimate stress increases with the mass volume of the concrete. The following expression describes this relation reasonably well:

$$\frac{f'_{b0}}{f'_b} = \left(\frac{A}{A_0}\right)^{\frac{\rho}{4800}}$$

in which:

ρ is the density of the concrete in kg/m^3.

3.2.3 Conclusions

The most important conclusions from this part of the investigation are:

- Although for this part of the investigation no numerical analysis was conducted, the conclusion can be drawn that the failure behaviour of gravel concrete and lightweight concrete differs when the load is distributed in two directions. This can probably be ascribed to the differences in the crack energy values.
- The influence of the dimensions of the loaded area on the ultimate stress under the loaded area is smaller with lightweight concrete than with gravel concrete. For gravel concrete the ultimate stress, in accordance with the relation given in

32

article 9.14.2 of the VCB 1990, is approximately proportional to $\sqrt{A/A_0}$. On the basis of the test results, the following relation can be used for lightweight concrete:

$$f'_{b0} = f'_b \left(\frac{A}{A_0} \right)^{\frac{\rho}{4800}}$$

in which:

ρ is the density of the concrete in kg/m^3.

- As is also valid for the relation for gravel concrete, the given relation is independent of the strength class of the concrete. However, the impression exists that this relation is not valid for strength classes above B 65.
- The failure mechanism of lightweight concrete is identical to that of gravel concrete. With large dimensions of the loaded area, shear of a corner of the prism occurs. With strongly concentrated loads, a conical wedge is formed that splits the concrete.

CHAPTER 4

BOND BEHAVIOUR

4.1   Tension bars

4.1.1 *General*

To gain insight into the crack behaviour of reinforced lightweight concrete a number of centrically reinforced tension bars was investigated both experimentally and numerically. In the experimental research the following material properties were determined: the uniaxial concrete tensile strength, the tension-softening and the bond behaviour. These data were subsequently used in the numerical research.

4.1.2 *Experimental research into the bond behaviour*

The experimental research [9] consisted of the following parts:
- determination of the uniaxial tensile strength and the tension-softening behav
  iour by means of deformation controlled tensile tests on plain concrete drilled
  cores;
- determination of the bond behaviour by means of pull-out tests with a bond
  length of thrice the diameter, for three different diameters (∅12, ∅16 and
  ∅20);
- research into the crack formation in reinforced tension bars, in which the
  reinforcement was varied (reinforcement 1 ∅12, 1 ∅16, 1 ∅20 and 4 ∅10 with
  ω_0 = 1.0 %, 2.0 %, 3.1 % and 3.1 %, respectively).

The experiments were executed with gravel concrete B 30 and B 60, Lytag concrete B 30 and B 60, Aardelite concrete B 30 and Liapor concrete B 30. Below each of the three parts of the investigation will be discussed in detail.

Tensile strength and tension-softening
The tensile tests were executed deformation controlled on drilled cores with a diameter of 64 mm, in which a notch had been cut all around with a depth of 5 mm. The elongations were measured over a measuring length of 110 mm. The tensile test is schematically represented in figure 17.
Figure 18 shows the behaviour of the various types of concrete, after the tensile strength has been reached. The ratio of tensile stress and the tensile strength is plotted on the vertical axis and the elongation over the measuring length of 110 mm is plotted on the horizontal axis. From the surface under these curves the values of the crack energy, which are also given in the figure, were calculated. For gravel and lightweight concrete the following relations are approximately valid for G_f:

gravel concrete: $\quad G_f = 24 + 26f_b$ J/m^2 with f_b in N/mm^2
lightweight concrete: $\quad G_f = 24 + 16f_b$ J/m^2 with f_b in N/mm^2

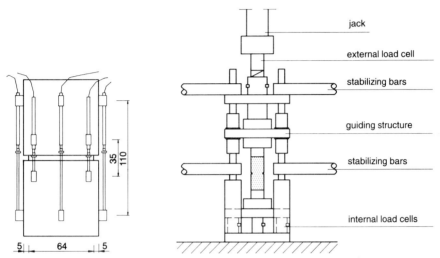

Fig. 17. Experimental research into prisms loaded in tension.

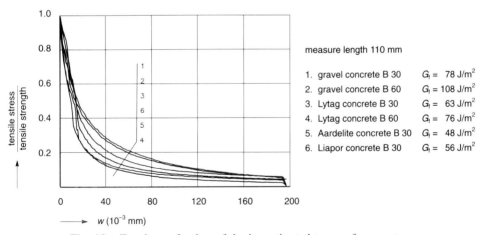

Fig. 18. Tension softening of the investigated types of concrete.

The relation for gravel concrete is based on a large number of research results [10]. For the results of the lightweight concrete investigated here the given expression is a reasonable approximation.

For the purpose of the numerical investigation the following formula was used to describe the softening behaviour [6]:

$$\frac{\sigma_b}{f_b} = \left\{1 + \left(c_1\frac{w}{w_0}\right)^3\right\} e^{-c_2\frac{w}{w_0}} - \frac{w}{w_0}(1 + c_1^3)e^{-c_2}$$

in which:

> w is the crack width;
> $w_0 = 5.14\ G_f/f_b$;
> $c_1 = 3$;
> $c_2 = 6.93$.

This formula provides a good approximation of the behaviour established in the tests.

Pull-out tests

The pull-out tests were executed on cube-shaped test specimen. The bond length of the reinforcement bar equals three times the bar diameter. Figure 19 shows a schematic representation of the pull-out test.

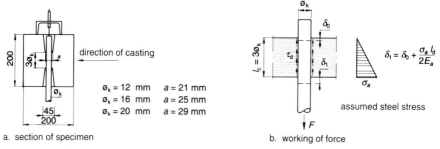

$\varnothing_k = 12$ mm $a = 21$ mm
$\varnothing_k = 16$ mm $a = 25$ mm
$\varnothing_k = 20$ mm $a = 29$ mm

$\delta_1 = \delta_0 + \frac{\sigma_a\,l_d}{2E_a}$

assumed steel stress

a. section of specimen   b. working of force

Fig. 19. Experimental research into the pull-out behaviour.

The following parameters were investigated:

- the aggregate (Lytag, Aardelite, Liapor, gravel);
- the strength classes B 30 and B 60;
- the bar diameter (∅12, ∅16 and ∅20).

Figure 20 summarizes the results of the pull-out tests. The values given in this figure are the average values per type of concrete. This approach is considered to be justified since, theoretically, the bar diameter does not influence the bond behaviour, as the concrete cover was sufficiently large to prevent splitting. Because the bond strength is usually regarded as a function of the tensile strength, the ratio of the bond stress and the tensile strength has been given in figure 20. The diagram shows that the relative bond strength of lightweight con-

36

crete is considerably lower than that of gravel concrete. The various types of lightweight concrete show comparable behaviour, despite the differences in density of the aggregates, which varied from 940 kg/m³ for Liapor concrete to 1630 kg/m³ for Aardelite concrete. It is concluded that the bond strength is determined to an important extent by the compressive strength of the particles.

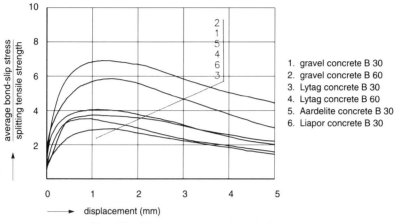

Fig. 20. Measured bond-slip relations.

Tension bars

Figure 21 shows the behaviour of a reinforced bar when a tensile test is carried out. As soon as the concrete tensile strength is reached in a cross section, a crack occurs and the tensile force is locally transferred by the reinforcement. If the load is increased, a second crack will appear at some distance from the first one, since a certain length is required before, through bond stresses, a sufficiently large tensile force has been built up in the concrete to create a new crack. This process will continue until the crack pattern is fully developed and no new cracks appear.

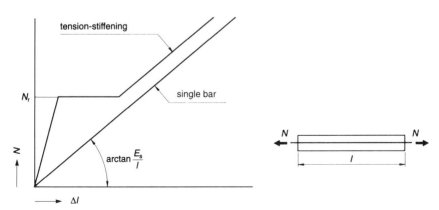

Fig. 21. Schematic representation of the force-elongation diagram of a reinforced tension bar.

37

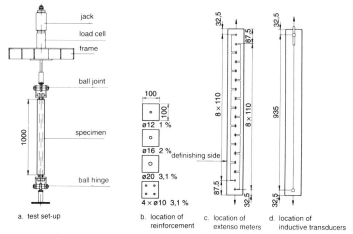

a. test set-up

b. location of reinforcement

c. location of extenso meters

d. location of inductive transducers

Fig. 22.   Experimental research into tension bars.

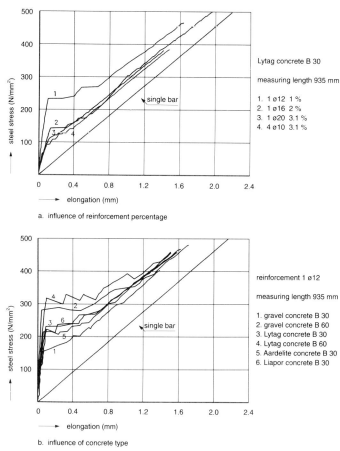

a. influence of reinforcement percentage

Lytag concrete B 30

measuring length 935 mm

1. 1 ø12 1 %
2. 1 ø16 2 %
3. 1 ø20 3.1 %
4. 4 ø10 3.1 %

b. influence of concrete type

reinforcement 1 ø12

measuring length 935 mm

1. gravel concrete B 30
2. gravel concrete B 60
3. Lytag concrete B 30
4. Lytag concrete B 60
5. Aardelite concrete B 30
6. Liapor concrete B 30

Fig. 23.   Elongation of the tension bar as a function of the steel stress in the crack.

Figure 21 shows that the stiffness of the tension bar is larger than that of a single reinforcement bar, also in the area where the crack pattern is fully developed. This increase in stiffness is called 'tension-stiffening' and it occurs because the concrete between the cracks continues to contribute to the load transfer.

The set-up of the executed tests and the position of the reinforcement are shown in figure 22. This figure also shows the measurements that were carried out. Figure 23 shows two representative examples of the recorded tension-deformation behaviour. The influence of the reinforcement percentage becomes clear from figure 23a. When the reinforcement percentage increases, the influence of tension-stiffening decreases. The reinforcement distribution (1 Ø20 or 4 Ø10) hardly plays a role. The influence of the type of concrete is shown in figure 23b. This influence is mainly determined by the differences in the concrete tensile strength. No significant differences in crack behaviour were found between light-weight concrete and gravel concrete with comparable tensile strengths.

4.1.3 Numerical research into the bond behaviour

For the purpose of the numerical research into the bond behaviour [11], the tension bar was modelled using axial symmetry. The circular section of the bar assumed in the calculations has the same cross-sectional area as the square cross section in the experiments. Between the reinforcement bar and the concrete, so-called interface elements were modelled, which made it possible to model the bond-slip characteristic. The analysis was carried out for tension bars with material properties as specified for Lytag concrete B 30 and B 60. Per strength class two bar diameters were investigated: Ø12 and Ø20.

The values of the modulus of elasticity, the tensile strength and the crack energy of the concrete were based on the experimental investigation. A summary is given in table 5. The values given for the tensile strength are average values since the tensile strength was entered as a stochastic variable, for which a normal distribution with a variation coefficient of 0.05 was used. This results in a localization of cracks. In each of the four cases the τ-Δ-relation, determined in the pull-out tests described above, was used as the bond-slip characteristic.

Table 5. Summary of the executed numerical simulations regarding the tension bars.

number	reinforcement	strength class	crack energy (J/m²)	tensile strength (N/mm²)	modulus of elasticity (N/mm²)
1	Ø12	B 30	63	1.96	20400
2	Ø12	B 60	76	3.49	26900
3	Ø20	B 30	63	1.96	20400
4	Ø20	B 60	76	3.49	26900

The calculation results for the strength class B 30 are shown in the figures 24 and 25. The distribution of the bond stress shows that only the initial stage of the τ-Δ-relation is covered. This means that hardly any slip occurs. Moreover, it can be concluded that the differences in τ-Δ-relations for gravel concrete and lightweight concrete, as established in the experimental research, hardly affect the behaviour of the tension bar, because these differences only become manifest much later in the τ-Δ-diagram. Therefore it is concludedë that the numerical and the experimental results correspond very well.

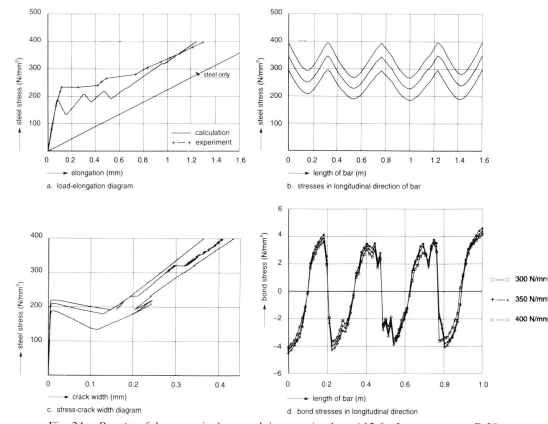

Fig. 24.   Results of the numerical research into tension bars Δ12 for Lytag concrete B 30.

To establish the effect of variations in the entered τ-Δ-relations, additional calculations were carried out. These variations were executed in comparison with the calculation with the material properties of Lytag concrete B 60, with one reinforcement bar ∅12. The characteristics of the various elasto-plastic relations are shown in table 6.

These calculations showed that only the cases with higher bond strengths led to results which are reasonably consistent with the tests. This means that the crack-

40

ing behaviour can be simulated reasonably well when a τ-Δ-relation is applied with a sufficiently high bond strength. This is caused by the fact that the behaviour in these cases is governed by the failure of the concrete directly adjacent to the reinforcement bar.

Table 6. Summary of executed parameter variations with regard to the tension bars.

number	grade τ/Δ (N/mm² · mm)	bond (N/mm²)
1	8000	2
2	8000	5
3	8000	8
4	200	2
5	200	5
6	200	8

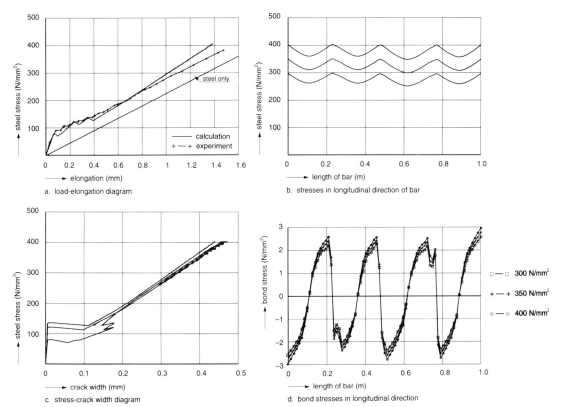

Fig. 25. Results of the numerical research into tension bars Ø20 for Lytag concrete B 30.

This means that the behaviour can be very well numerically predicted if complete bonding is assumed. This, however, requires a fine element mesh, as used in these calculations. The input of a characteristic, as found in a pull-out test, does not appear to be necessary, since only the initial stage of this characteristic is covered.

4.1.4 Conclusions

The most important conclusions to be drawn from the investigation into bond behaviour are:

- The tension-softening behaviour of the investigated lightweight concrete does not differ substantially from the behaviour of gravel concrete. The crack energy of lightweight concrete, however, is 20 to 30 % lower.
- The bond strength of lightweight concrete, as established in a pull-out test, is considerably lower than the bond strength of gravel concrete with the same splitting tensile strength.
- The crack behaviour of lightweight concrete does not differ from the crack behaviour of gravel concrete. This is valid for the crack distance, the crack width and the tension-stiffening.
- The differences between lightweight concrete and gravel concrete found in the pull-out test, have no influence on the crack behaviour. This is due to the fact that only the initial stage of the τ-Δ-relation, in which there is no essential difference, is covered.
- The crack behaviour can very well be simulated numerically. For this purpose it is sufficient to model complete bonding, in combination with a fine element mesh. The crack behaviour of tension bars made of deviating types of lightweight concrete can therefore be predicted, provided that the concrete properties (modulus of elasticity, tensile strength and crack energy) are known.

4.2 Laps

4.2.1 Experimental research

The objective of the investigation concerning laps [12] was to establish possible differences between lightweight concrete and gravel concrete. Only experimental research was carried out. Tests were carried out on beams with a cross section of 400 mm × 300 mm that were subjected to a four-point bending test (see figure 26). At the laps the beams were provided with transverse reinforcement, which was anchored by means of welded plates. In this way the situation in a slab, in which confinement in horizontal direction occurs, was simulated.

In all 13 tests were carried out, with the following variables:

- the coarse aggregate (gravel, Lytag, Liapor, Aardelite);
- the reinforcement percentage, together with the concrete cover:
  4 $\varnothing 20$ with $\omega_0 = 1.21$ % with a 30 mm concrete cover;
  3 $\varnothing 20$ with $\omega_0 = 1.02$ % with a 60 mm concrete cover;
- the length of the laps:
  Liapor concrete $l_1 = 20\varnothing_k$, $30\varnothing_k$, $40\varnothing_k$ and $50\varnothing_k$;
  other tests $l_1 = 30\varnothing_k$.

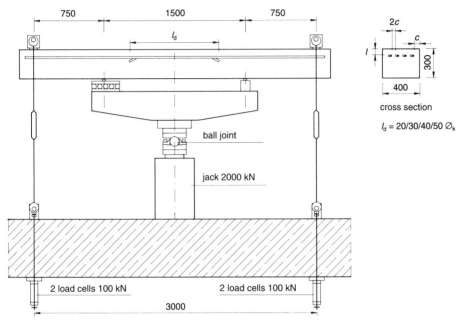

Fig. 26. Experimental research of laps.

All bars were provided with laps at the same position, with the exception of test beam 13, which contained two continuous bars. Table 7 contains a summary of the executed tests and their results.

Table 7. Summary of the test results of the research into laps.

test	aggregate	lap length (mm)	F_u (kN)	M_u (kNm)	$M_{u\,ber}$ (kNm)	σ_s (N/mm²)	τ_u (N/mm²)	f_b (N/mm²)	τ_u/f_b
reinforcement 4∅20 – concrete cover 30 mm									
1	Liapor	400	97.1	72.8	140.6	257	3.21	2.32	1.38
2	Liapor	600	139.7	104.8	139.2	371	3.10	1.99	1.55
3	Liapor	800	156.6	117.5	140.4	416	2.60	2.42	1.07
4	Liapor	1000	179.2	134.4	140.1	479	2.40	2.23	1.08
8	Lytag	600	133.4	100.1	133.4	354	2.95	2.11	1.40
9	Aardelite	600	149.3	112.0	134.4	400	3.33	1.96	1.70
11	gravel	600	158.8	119.1	134.8	424	3.54	1.96	1.80
13	gravel	600	174.0	130.5	139.5	459	3.82	2.39	1.60
14	gravel	600	153.5	115.1	135.8	406	3.38	2.03	1.67
reinforcement 3 ∅20 – concrete cover 60 mm									
6	Liapor	600	110.2	82.6	94.7	437	3.64	2.11	1.72
7	Lytag	600	94.0	70.5	91.1	373	3.11	1.85	1.68
10	Aardelite	600	106.5	79.9	93.8	419	3.49	1.97	1.77
12	gravel	600	122.2	91.6	92.2	490	4.08	1.96	2.08

The results in this table concern:

- the ultimate load and the bending moment of failure;
- the calculated bending moment of failure;
- the steel stress at the ultimate load;
- the (average) bond strength and the splitting tensile strength;
- the ratio of the bond strength and the splitting tensile strength.

All test beams failed due to failure of the laps. Figure 27 shows a typical failure mode. The test results are given in figure 28. The figures 28a and 28b make clear that the failure load does not increase proportionally with the lap length. It should be noted, however, that the laps investigated here are not only loaded in tension, but also bent. Bending causes additional splitting stresses perpendicularly to the surface of the reinforcement, which adversely influence the strength.

Figure 29 shows the maximum obtained bond stress, divided by the splitting tensile strength. It appears that the laps in lightweight concrete systematically have a lower relative strength than the laps in gravel concrete. Possibly the lower crack energy of lightweight concrete plays a role in this. No explanation can be given for the differences between the various lightweight aggregates. Possibly these differences are due to scatter in the test results. On average, the bond strength of lightweight concrete is approximately 15 % lower than the bond strength of gravel concrete. Since the strength of the laps increases less than proportionally to the lap length, it is advisable to extend the lap for lightweight concrete with about 30% to obtain the same strength as for gravel concrete.

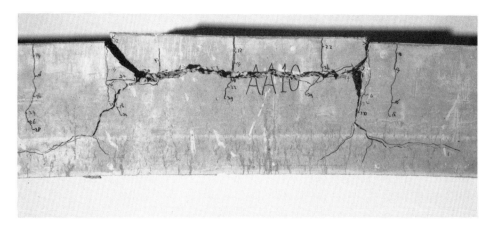

Fig. 27. Characteristic failure mode of laps.

All the tests executed in this investigation resulted in an explosive failure mode. Immediately after the appearance of horizontal splitting cracks at the level of the reinforcement the concrete cover burst off completely, causing the reinforcement to come loose, which in turn made load transfer impossible. This shows once more the importance of a good detailing. Preferably, the laps should not be positioned at critical points in a structure. Moreover, the laps should be staggered as far as possible.

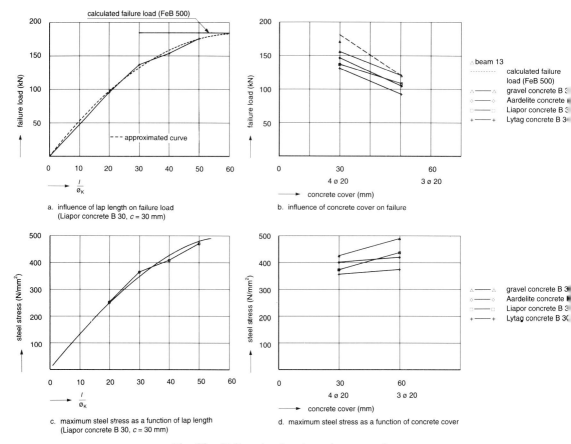

a. influence of lap length on failure load
(Liapor concrete B 30, c = 30 mm)

b. influence of concrete cover on failure

c. maximum steel stress as a function of lap length
(Liapor concrete B 30, c = 30 mm)

d. maximum steel stress as a function of concrete cover

Fig. 28.   Failure load and maximum steel stress.

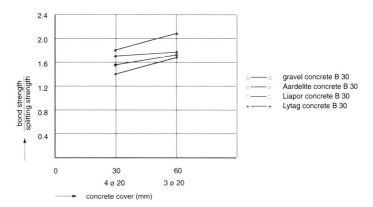

Fig. 29.   Relative bond strength as a function of the concrete cover.

46

4.2.2 *Conclusions*

The most important conclusion from the investigation regarding laps are:

- The strength of the laps is not proportional to the lap length.
- The bond strength of lightweight concrete is about 15 % lower than that of gravel concrete of the same strength class. Also with regard to the first conclusion the lap length for lightweight concrete has to be increased by 30 % as compared to the value for gravel concrete.
- The failure behaviour of lightweight concrete beams does not differ essentially from the behaviour of gravel concrete. When the laps are not arranged staggered, the failure of the lap will cause sudden, brittle fracture.
- For large diameter bars and/or bars that are situated closely together, which have laps in the same cross section, it is advisable to define more stringent requirements with regard to the detailing by increasing the minimal distance between the laps.

CHAPTER 5

SHEAR BEHAVIOUR

5.1 General

The usual dimensioning of beams for shear is based on a large number of research results, in combination with generally accepted theories to describe the various mechanisms that play a role in shear behaviour. In comparison with gravel concrete, however, little information appeared to be available on the shear strength of light weight concrete beams with shear reinforcement.

Since the deviating crack behaviour of lightweight concrete, in which the cracks run through the aggregate particles instead of around them, may influence the shear behaviour, both experimental and numerical research was carried out. The research was carried out on beams of reinforced concrete with an I-shaped cross section. The attention was mainly focused on the aggregate interlock in the cracks and the rotation of the diagonal compression struts during the load increase on the beams.

5.2 Experimental research

The experimental research [13] included five series with different types of concrete of three test specimen each, in which the amount of shear reinforcement was varied per test specimen. The shear reinforcement consisted in all cases of stirrups. A summary of the executed tests and their results is given in table 8. The variables were:

- the coarse aggregate (gravel, Lytag, Aardelite, Liapor);
- the strength class (Lytag concrete only, B 30 and B 60);
- the amount of shear reinforcement.

Figure 30 shows the test set-up. During the tests the following was measured:

- the elongations and shortenings in the areas with shear transfer at the web of the beam with different measuring grids;
- the elongations and shortenings in the areas with shear transfer at the top of the top flange and at the underside of the bottom flange;
- the deflection of the beam in the centre of the span and at the two point loads.

On the basis of the measurements at the web of the beams the development of the shear deformation and the inclination of the compression struts were calculated, both as a function of the load. Also the normal and parallel displacements in the diagonal cracks in the areas with shear transfer were determined.

Table 8. Summary of the test results with regard to shear behaviour.

beam	aggregate	shear reinforce-ment (%)	strength class	f'_{ck} (N/mm²)	f_b (N/mm²)	V_u (kN)	τ_{exp} * (kN)	failure mode
gd30l	gravel	0.430	B 30	28.40	2.27	359.5	4.73	yielding
gd30m	gravel	0.887	B 30	22.00	1.82	420.0	5.53	crushing
gd30h	gravel	1.450	B 30	30.90	2.46	470.0	6.35	crushing
Lg30l	Lytag	0.430	B 30	23.90	1.68	324.0	4.26	yielding
Lg30m	Lytag	0.887	B 30	35.50	2.64	520.0	6.86	crushing
Lg30h	Lytag	1.450	B 30	31.50	2.43	481.5	6.34	crushing
Lr30l	Liapor	0.430	B 30	34.20	2.44	330.0	4.34	yielding
Lr30m	Liapor	0.887	B 30	31.30	2.34	461.0	6.07	crushing
Lr30h	Liapor	1.450	B 30	31.50	2.32	541.0	7.31	crushing
Ae30l	Aardelite	0.430	B 30	28.50	2.06	321.5	4.23	yielding
Ae30m	Aardelite	0.887	B 30	27.10	2.08	457.5	6.02	crushing
Ae30h	Aardelite	1.450	B 30	25.30	1.95	482.0	6.51	crushing
Lg60l	Lytag	0.660	B 60	54.20	2.96	517.0	6.80	yielding
Lg60m	Lytag	1.250	B 60	57.90	3.17	751.0	10.75	crushing
Lg60h	Lytag	2.700	B 60	57.10	2.84	881.0	11.90	crushing

$$* \ \tau_{exp} = \frac{V_u}{b_w d}$$

in which:

b_w is the web thickness;

d is the effective height of the beam with : $d = 760$ mm when bottom reinforcement in 1 layer and $d = 740$ mm when bottom reinforcement in 2 layers.

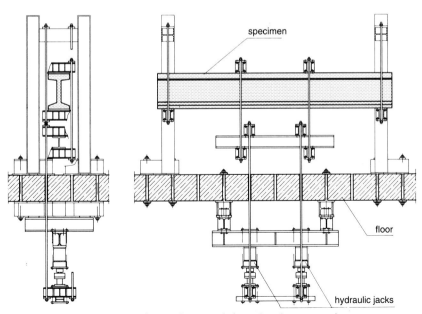

Fig. 30 Experimental research into the shear capacity.

Table 8 contains a summary of the registered failure load and the failure mode of the various beams. Also the corresponding values of the cube compressive strength and the splitting tensile strength are given in this table. With the beams with a small shear reinforcement percentage the failure is always introduced by yielding of the shear reinforcement. With the other beams this stage was never reached, since crushing of the diagonal compression struts turned out to be the determining factor.

Figure 31 shows the measured deflection as a function of the load on the beams for the three Liapor concrete B 30 beams. The beam with the highest percentage of shear reinforcement shows a higher cracked stiffness than the other two beams. This is caused by the higher percentage of longitudinal reinforcement in the beams with the highest percentage shear reinforcement. With regard to the failure behaviour this figure shows that this behaviour becomes more brittle as the amount of transverse reinforcement increases. This is similar for the other tested beams.

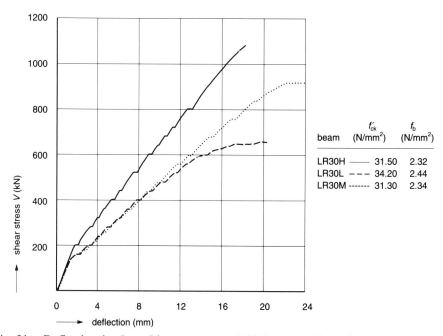

Fig. 31. Deflection in three Liapor concrete B30 beams with various percentages of shear reinforcement.

The gradient of the diagonal compression struts, which is initially 45° in the uncracked stage, turns out to be less in the failure stage. The lower the shear reinforcement percentage, the higher the decrease. No significant differences between lightweight concrete and gravel concrete were found. This means that in spite of the fact that the cracks in lightweight concrete run through the aggregate particles, a shear stress comparable to that of gravel concrete is still transferred.

50

This is probably due to the somewhat undulating shape of the crack surfaces, which still allows the formation of contact points after parallel displacements of the crack surfaces.

This is also valid for the observed crack gradients. The gradients of the cracks that were formed in the successive stages of loading decreased more in the beams with a low percentage of shear reinforcement. The number of cracks appeared to increase as the amount of shear reinforcement was higher. It should be noted that in beams with a substantially slighter crack gradient than 45°, a crack with a steeper gradient that was formed earlier appeared to be the critical crack at failure. This seemed to occur more frequently with lightweight concrete beams.

Finally, the registered failure loads were compared to the values for gravel concrete beams calculated according to the VBC 1990. In the VBC 1990 a distinction is made between failure due to yielding of the shear reinforcement and due to failure of the diagonal compression struts. In the first case the value of $\tau_1 + \tau_s$ determines the shear resistance, in the second case it is the value of τ_2. The lower of these two values is governing. A summary of the results is given in table 9. A crack gradient of 45° was assumed.

Table 9. Ultimate shear strength at an assumed crack gradient $\theta = 45°$.

beam	f'_{ck} (N/mm^2)	τ_{exp} (N/mm^2)	τ_1 (N/mm^2)	τ_s (N/mm^2)	$\tau_1 + \tau_s$ (N/mm^2)	τ_2 (N/mm^2)	τ_{VBC} (N/mm^2)	τ_{exp}/τ_{VBC} (N/mm^2)
gd30l	28.40	4.73	1.64	2.15	3.79	4.79	3.79	1.25
gd30m	22.00	5.53	1.43	4.86	6.29	3.74	3.74	1.48
gd30h	30.90	6.35	1.73	7.96	9.69	5.25	5.25	1.21
Lg30l	23.90	4.26	1.49	2.15	3.64	4.06	3.62	1.17
Lg30m	35.50	6.84	1.88	4.86	6.74	6.04	6.04	1.13
Lg30h	31.50	7.31	1.75	7.96	9.71	5.36	5.36	1.18
Lr30l	34.20	4.34	1.83	2.15	3.98	5.81	3.98	1.09
Lr30m	31.30	6.07	1.74	4.86	6.60	5.32	5.32	1.14
Lr30h	31.50	7.31	1.75	7.96	9.71	5.36	5.36	1.36
Ae30l	28.50	4.23	1.65	2.15	3.80	4.85	3.80	1.11
Ae30m	27.10	6.02	1.60	4.86	6.46	4.61	4.61	1.31
Ae30h	25.30	6.51	1.54	7.96	9.50	4.30	4.30	1.51
Lg60l	54.20	6.80	2.50	3.61	6.11	7.00	6.11	1.11
Lg60m	57.90	10.75	2.62	6.82	9.44	7.00	7.00	1.54
Lg60h	57.10	11.90	2.60	13.57	16.17	7.00	7.00	1.70

A similar summary is given in table 10, with an assumed crack gradient of 30°. When a crack gradient of 45° is assumed, all test results appear to be higher than predicted by the VBC 1990. However, the strength of the lightweight concrete beams turns out to be somewhat lower than that of the gravel concrete beams. The average difference is 7 %.

When a crack gradient of 30° is assumed, the value of τ_2 is governing in practically all cases, including those with a lower percentage of shear reinforcement. This is in contradiction with the observed failure modes, in which ultimately a steeper crack gradient and, consequently, the yielding of the shear reinforcement turned out to be the failure mode. As is also apparent from table 10, allowing such low crack gradients in the calculation leads to overestimation of the strength. Because of this the crack gradient needs to be limited, especially for lightweight concrete.

Table 10. Ultimate shear stress at an assumed crack gradient of $\theta = 30°$.

beam	f'_{ck} (N/mm²)	τ_{exp} (N/mm²)	τ_1 (N/mm²)	τ_s (N/mm²)	$\tau_1 + \tau_s$ (N/mm²)	τ_2 (N/mm²)	τ_{VBC} (N/mm²)	τ_{exp}/τ_{VBC} (N/mm²)
gd30l	28.40	4.73	1.64	3.72	5.36	4.83	4.83	0.98
gd30m	22.00	5.53	1.43	8.41	9.84	3.74	3.74	1.48
gd30h	30.90	6.35	1.73	13.77	15.50	5.25	5.25	1.21
Lg30l	23.90	4.26	1.49	3.72	5.21	4.06	4.06	1.17
Lg30m	35.50	6.84	1.88	8.41	10.29	6.04	6.04	1.13
Lg30h	31.50	7.31	1.75	13.77	15.52	5.36	5.36	1.18
Lr30l	34.20	4.34	1.83	3.72	5.55	5.81	5.55	0.78
Lr30m	31.30	6.07	1.74	8.41	10.15	5.32	5.32	1.14
Lr30h	31.50	7.31	1.75	13.77	15.52	5.36	5.36	1.36
Ae30l	28.50	4.23	1.65	3.72	5.37	4.85	4.85	0.87
Ae30m	27.10	6.02	1.60	8.41	10.01	4.61	4.61	1.31
Ae30h	25.30	6.51	1.54	13.77	15.31	4.30	4.30	1.51
Lg60l	54.20	6.80	2.50	6.25	8.75	7.00	7.00	0.97
Lg60m	57.90	10.75	2.62	11.80	14.42	7.00	7.00	1.54
Lg60h	57.10	11.90	2.60	23.18	26.08	7.00	7.00	1.70

5.3 Numerical research

Within the framework of the numerical research both simulation calculations [14] and parameter analyses [15] were carried out. The simulation calculations concern three beams with material properties and geometry corresponding to the tested Lytag concrete B 30 beams. In the modelling symmetry in relation to the centre of the span was used. As a function of an increasing load were determined:

- the deflections in the centre of the beam of and below the point load;
- the elongations at the measuring positions used in the experiments;
- the development of the concrete compressive strain;

52

- the longitudinal distribution and the development of the maximum concrete compressive stress, concrete compressive strain and stirrup stresses.
- the distribution and the development of the tensile stress in the longitudinal reinforcement at the tension side;
- the crack formation.

Figure 32 shows the calculated and the measured load-deflection diagrams for the centre of the span. It turns out that the stiffness of the beams at higher loads is somewhat overestimated in the calculation. Figure 33 shows the development of the stirrup stress during the successive loading steps. This figure clearly shows that in beam 1 the yield stress of 580 N/mm² is reached, whereas this is not the case in the beams 2 and 3. Figure 34 shows the development of the stress in the longitudinal reinforcement. It clearly shows that the steel stress in the area between the support and the point load is higher than follows from the moment and the internal lever arms. The shift length is more ore less equal to half the beam hight and is higher in the beam with the low percentage of shear reinforce-ment, although the differences are only small. In figure 35 the crack patterns as found in the calculations have been combined with the cracks as observed in the experiments. Here, too, there is a reasonable correspondence.
For the somewhat higher stiffness, which was found numerically, three possible causes can be indicated:

a. Concrete in compression is modelled elasto-plastic. Because of this the deformations under higher stresses are underestimated, whereas, because no 'softening' has been modelled, the stress is overestimated with high com-pressive strains.
b. Though the model that describes the "aggregate interlock" in the cracks results in a smaller increase of shear stress at increasing crack strain, the absolute value of the shear stress continues to increase.
c. When in the calculation a crack is formed at a certain point, a new crack can only develop at that same point if the gradient deviates more than 60° from the gradient of the first crack.

The points mentioned above also cause that numerically no top was found in the load-deflection diagram. The criterion adhered to in defining failure in the calcu-lation is that a concrete compressive strain of approximately 3 ‰ is reached. As a result a numerical failure load was established for all cases that deviated less than 10 % from the values that were found experimentally. The same result would have been found when a limit value for the compressive strain of 2.5 ‰ and 4 ‰ had been used. Within this range the effect on the failure load is apparently small.

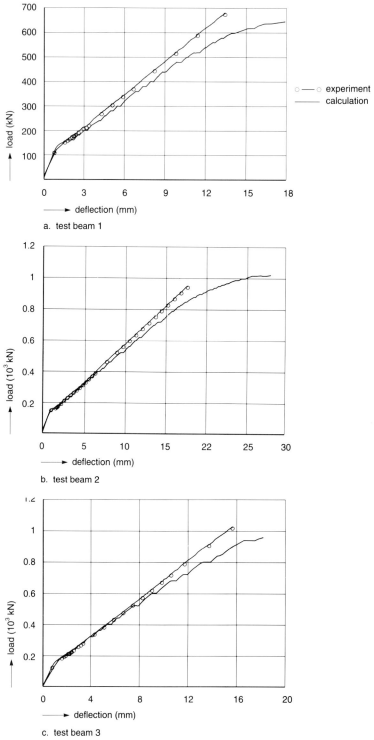

Fig. 32. Calculated and measured load-deflection diagrams.

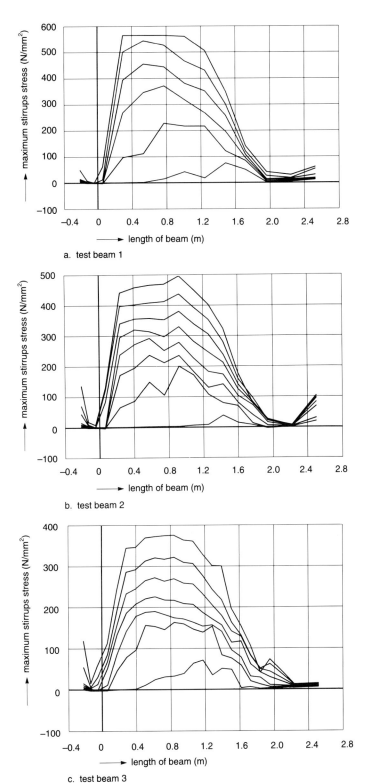

a. test beam 1

b. test beam 2

c. test beam 3

Fig. 33. Development of the stirrup stresses.

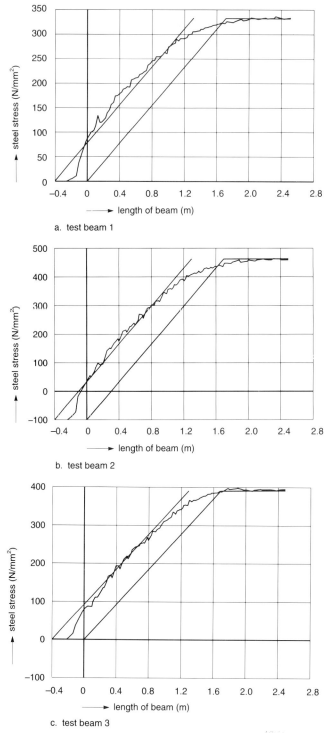

a. test beam 1

b. test beam 2

c. test beam 3

Fig. 34. Development of the stress in the longitudinal reinforcement.

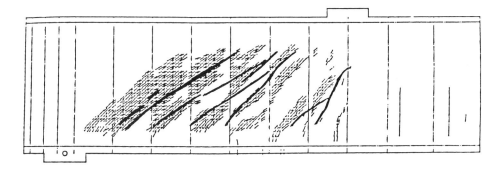

a. test beam 1

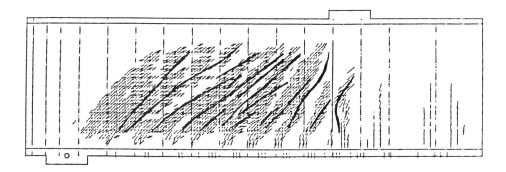

b. test beam 2

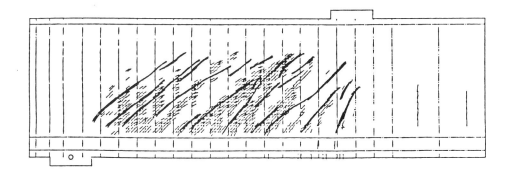

c. test beam 3

Fig. 35. Comparison of the calculated and the observed crack patterns.

With the beam with the lowest percentage of transverse reinforcement as reference, three different variations were introduced in the parameter analysis. These concerned:

- variation 1: a substantially reduced "aggregate-interlock" factor;
- variation 2: a reduction of the crack energy with approximately 50 %;
- variation 3: a curved ascending branch in the σ-ε-diagram of concrete in compression.

Figure 36 shows the load-deflection diagrams for the centre of the beams for these variations. Variation 1 appears to lead to an important improvement. It should be noted that the reduction of the "aggregate-interlock" was also taken into account at lower load levels, which is actually not justified in that area, since the low crack strains in this area allow the development of higher shear stresses. Variation 2, with reduced crack energy, only has influence at lower load levels. This influence, however, is small. The third variation, in which the ascending branch of the σ-ε-diagram of concrete in compression has been modelled more realistically, also appears to be of little influence.

It is concluded that it is possible to describe the beam behaviour numerically reasonably well with the presently available modelling of the material behaviour. The shortcomings in this modelling are in particular the shear behaviour in the crack (no decrease of shear stress after the initial increase at increasing crack strain) and the behaviour of concrete in compression (no "softening"). On the basis of an arbitrary value of the compressive strain of 3 ‰, the experimental failure load can be approximated reasonably well by using the same basic properties that were used in the reference calculation. The result is only slightly affected by the shape of the ascending branch of the σ-ε-diagram for concrete in compression and the crack energy. It may be expected that a further re-distribution of concrete compressive stresses will occur when "softening" is modelled under pressure. Because of this a further decrease of the stiffness and, moreover, a top in the load-deflections diagram will be found, so that also in the calculation the failure load can be indicated unambiguously.

Finally, on the basis of the numerical analysis it can be stated that the differences in shear behaviour between beams with material models that correspond with gravel concrete and lightweight concrete are only small. The most important difference is caused by the lower modulus of elasticity, the effect of which can be determined very well in the numerical simulation. The effect of the deviating values of the crack energy, as occurs with lightweight concrete, appears to be small. Also the effect of variations in the amount of shear reinforcement within the investigated area can be simulated well. So it is possible to numerically predict the shear behaviour in this area provided the basic material properties are known. In this way the shear behaviour of other types of lightweight concrete can be investigated numerically when the material properties have been established experimentally. However, with lower amounts of shear reinforcement than applied here, it is expected that the numerical simulation will become difficult because of numerical instabilities. Extrapolation to lower amounts of shear reinforcement is therefore not very well possible.

58

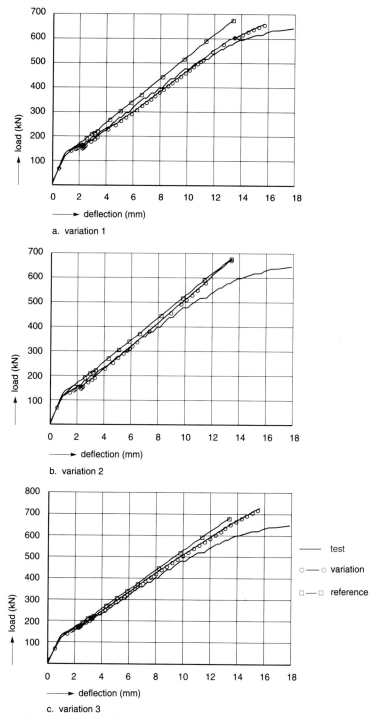

a. variation 1

b. variation 2

c. variation 3

Fig. 36. Load-deflection diagrams from the parameter analysis.

5.4 Conclusions

The most important conclusions of the investigation into shear behaviour are:

- in relation to the shear strength according to the VBC 1990, the shear strength of lightweight concrete beams with an average cube compressive strength of 30 N/mm^2 is on average 7% lower than the shear strength of gravel concrete with an identical cube compressive strength;
- in all cases the measured shear strength was higher than the value according to the VBC 1990 under the assumption that the diagonal compression strut forms an angle of 45° with the axis of the beam;
- an assumed angle for the diagonal compression strut of 30° leads to overestimation of the bearing capacity; a limitation of this angle is therefore necessary;
- the beams with a low percentage of shear reinforcement failed due to yielding of the shear reinforcement; the beams with a medium or high percentage of reinforcement failed because of failure of the diagonal compression struts;
- it can be concluded from the rotation of the diagonal compression struts in lightweight concrete that the cracks transfer considerable shear stresses, despite the fact that these cracks run through the aggregate particles; a possible explanation is the irregular form of the crack surface which causes contact points to occur between the crack surfaces after parallel displacement;
- with the help of the presently available numerical possibilities it is possible to predict beam behaviour numerically, using a number of basic material properties;
- on the basis of a compressive strain criterion for concrete of about 3 ‰, the shear strength can be numerically predicted fairly accurately.
- the shear behaviour of beams made of types of concrete with other lightweight aggregates can be numerically determined fairly well when the material properties are known.

CHAPTER 6

DURABILITY

6.1 Influence of the type of aggregate particles on the corrosion process

6.1.1 *General*

The durability of reinforced or pre-stressed concrete is mainly determined by the extent to which the steel can corrode. This corrosion process is mainly based on the possible occurrence of transport phenomena. A distinction is made between transport through the cracks and transport through the uncracked concrete cover. In the following discussion both cases are examined further. Particular attention is paid to the role of the type of aggregate particles.

6.1.2 *Transport through cracks*

From a physical point of view cracks in lightweight concrete generally differ from cracks in gravel concrete. Cracks in gravel concrete occur through the cement matrix, along the edges of the aggregate particles. In lightweight concrete the cracks run through both the cement matrix and the lightweight aggregate particles. The differences in crack morphology are of importance in relation with the possibility to halt the corrosion mechanism by "repassifying" the steel. This repassifying is mainly due to "blockage" of the crack by corrosion products. This is valid both for corrosion by carbonation and corrosion by chloride penetration. Cracks in lightweight concrete occur through the aggregate particles. When an aggregate particle is situated near the reinforcement, such cracks may be interpreted as cracks that divide into a number of branches at the location of an aggregate particle, but their cross sections are, however, considerably smaller than the cross sections of the original cracks. This is shown in figure 37.

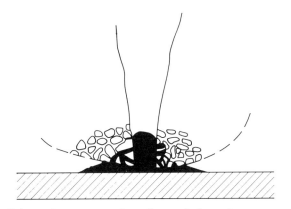

Fig. 37.   Schematic representation of a crack through a porous particle.

In this case the only difference that occurs between lightweight concrete and gravel concrete is the circumstance that in lightweight concrete more corrosion products have to be formed to halt the corrosion process. It may therefore be stated that the influence of porous aggregates on the corrosion process at cracks will not significantly differ from that in gravel concrete.

6.1.3 Transport through the concrete cover

The most important factor determining the corrosion process is the permeability of the concrete cover layer. This raises the question to what extent the porous lightweight aggregate particles contribute to the permeability of the concrete cover layer.

In [16] it is stated that the thickness $d*$, which determines the depth to which variations in the moisture content at the outside surface continues to affect the material, is characteristic for the dynamic-hygroscopic behaviour of the concrete cover layer. The value depends on two material properties namely: the moisture resistance number and the hygroscopic capacity. Table 11 shows the relative influence of a number of aggregates on the characteristic thickness $d*$.

Table 11. Relative influence of a number of aggregates on the characteristic thickness $d*$.

aggregate	relative μ-value	influence hygroscopy	influence affecting $d*$
pc:			
gravel	1.00	1.00	1.00
Lytag	0.84	2.00	0.77
Liapor	0.89	1.10	1.01
Lava	0.86	1.20	0.98
bfc:			
gravel	1.00	1.00	1.00
Lytag	0.65	1.20	1.13
Liapor	0.68	1.10	1.16
Lava	0.73	1.20	1.07

On the basis of these data the following conclusions can be drawn:

- With lightweight concrete mixtures based on portland cement the characteristic value of $d*$ remains the same or decreases (Lytag). A decrease in $d*$ means that changes in the moisture content will occur less deep in the surface layer in comparison with gravel concrete. As a result, the risks of reinforcement corrosion become less.
- With mixtures based on blast-furnace cement $d*$ increases with 10 to 15 %. This means that changes in the moisture content will manifest deeper in the

surface layer. With a concrete cover identical to that of gravel concrete, the risk of corrosion will increase. However, since the regulations require the concrete cover for lightweight concrete to be 5 mm thicker than that of gravel concrete, it can be stated that this extra concrete cover will be sufficient to compensate the influence of lightweight concrete mixtures on the dynamic-hygroscopical behaviour.

It may seem surprising that the differences in permeability between gravel concrete and lightweight concrete, with its porous particles, are so small: one would expect the particle porosity, "added" to the matrix porosity, to lead to an increase in the permeability.

Various investigations have shown that this interpretation is too simple. HOLM, BREMNER and NEWMAN [17], who executed research into lightweight concrete bridges in the United States of America, established on the basis of microscopic research into drilled cores that there is an exceptionally good adhesion between the lightweight aggregate particles and the matrix. Furthermore, they established the presence of an intermediate layer in the contact zone between aggregate particles and matrix with a thickness of approximately 60 µm, which clearly contained less pores and was much more homogeneous than the matrix at some distance from the particle surface.

For this conclusion, based on drilled cores from bridge decks of 20 to 40 years old that had been exposed to extreme circumstances, they found a conformation in Russian literature.

KHOKHRIN [18] stated that the adhesion of expanded aggregate particles to the matrix had been the subject of a number of Russian investigations. These studies led to the conclusion that the adhesion between expanded particles and matrix has a mechanical and a chemical side. In other words, the contact layer is not only a dividing layer between two substances but also contains new substances, which are formed as a result of the chemical interaction between the cement paste and the expanded aggregate particles. An increase in the cohesion is the result of the chemical reaction between the product of the cement hydration and the aluminium silicates on the surface of the lightweight aggregate particles, formed during the high production temperatures.

KHOKHRIN carried out measurements on the micro-hardness of the contact layer and the matrix at some distance of the particles, both for lightweight concrete and for gravel concrete. In the concrete with the expanded aggregate particles the hardness of the contact zone was much higher than that of the matrix at some distance from the particle surface. With gravel concrete no differences were found. As to the conclusions of this detailed research, KHOKHRIN points out that "the quality of the contact zone with regard to cohesion, density and strength is better with porous aggregates than with concrete with dense aggregates".

Another important aspect is the elastic compatibility. The most important reason for the less frequent occurrence of micro-cracks between lightweight aggregate particles and matrix are the more or less comparable elastic properties of the lightweight aggregate particles and the matrix. Stress-strain relations of light-weight concrete are characterized by linear behaviour up to about 90 % of the strength, which indicates strongly reduced micro cracking. When micro-cracking is limited the detrimental effect of micro-cracks filled with water in freezing cir-cumstances is also limited. At the same time there is no danger of fast penetration of chlorides. When there is no capillary continuity, chloride-ions have to diffuse through the cement paste (a slow process), which considerably slows down the building up of a higher concentration of chloride-ions at the reinforcement. In gravel concrete the stiffness of the particles is two to six times higher than the stiffness of the matrix. The use of large, stiff, dense aggregate particles in con-crete mixtures, with on top of that a higher air content, leads to high stress con-centrations in the contact surface between particles and matrix.

The possibility to extend the durability of concrete by reducing the stress concen-trations in the internal of the concrete by using expanded aggregate particles with a lower stiffness was also mentioned by PODVALNYI [19]. He concluded that "when the stresses in the conglomerate become smaller, the frost-thaw resistance becomes higher, so that lightweight concrete in this respect may even have a higher resistance than gravel concrete". KHOKHRIN [18] reported on various Rus-sian investigations, which showed that lightweight concrete has an equal, if not smaller permeability than gravel concrete. He attributed this phenomenon to the combination of an improved contact zone, a more homogeneous structure and a reduction of the internal stresses.

NISHI et al. [20] state that lightweight concrete has a higher resistance to penetra-tion of both fresh water and salt water than gravel concrete. They presume that this is a consequence of the forming of an isolating layer of very dense cement stone around the lightweight aggregate particles. HAYNES [21] found a more or less similar permeability in tests on lightweight concrete and gravel concrete.

In conclusion it can be stated that on the basis of the physical behaviour of light-weight concrete there is no reason to presume that this material is more sensitive to corrosion than gravel concrete.

6.2   Results of experimental research and practical observations

6.2.1 *General*

In 1986 the FIP-study group "lightweight concrete" presented its final report with findings concerning the various aspects of the behaviour of lightweight con-crete [22] at the FIP-congress in New-Delhi. Also published in 1986 was the Guide to the structural use of lightweight aggregate concrete of the Institution of Structural Engineers in the United Kingdom. Some of the conclusions mentioned in these reports are mentioned here.

6.2.2 *Resistance to frost-thaw cycles*

Important experimental information has been reported by WALSH [24]. Tests, carried out in 1959, demonstrated that concrete made with expanded soft slate has a higher resistance to frost-thaw cycles than gravel concrete with the same strength. Subsequently, tests were carried out on concrete slabs made with different mixtures. After three years of exposure, the test slabs had been exposed to 202 frost-thaw cycles and a 123-fold use of a salt solution. Of the 17 lightweight concrete slabs, 16 slabs appeared to be unaffected; 1 slab was slightly affected, which remained unchanged. Of the 27 gravel concrete slabs, only 4 slabs appeared to be damaged. Subsequent to these test series, in situ-tests were carried out on lightweight concrete bridge decks. These tests confirmed the picture formed by the tests carried out before.

Similar results have been reported by HOLM, BREMNER and NEWMAN [17] on the basis of observations on drilled cores, taken from lightweight concrete bridge decks. These bridge decks had been exposed to severe weather conditions and thaw salts without being damaged. Furthermore, five bridges in Japan, built in the period of 1967 to 1969, were inspected in 1983 [25]. Again the weather conditions were extremely unfavourable. The concrete strength class was relatively low (cylinder strengths of 21 to 24 N/mm²). Although two bridges showed some slight damage, no significant effect of the extreme cold could be noted. This is why in [22] it has been concluded that high-quality lightweight concrete, as is generally required for bridges, is also very durable under adverse weather conditions, even significantly more durable than gravel concrete. In [23] the use of air-entraining agents for these situations is expressly recommended.

6.2.3 *Carbonation*

The compactness of the concrete cover plays an important role as far as the danger of carbonation is concerned. Since the compactness of gravel concrete and lightweight concrete does not differ considerably, provided that the water-cement ratio is not too high and the amount of cement is sufficient, lightweight concrete may be expected not to behave worse than gravel concrete with regard to carbonation. This is confirmed by literature. OHUCHI et al. [25] reported on the condition of a railway platform that had been used for more than nineteen years. In the parts made with lightweight concrete a carbonation depth of 10 to 30 mm was found. In the parts made with gravel concrete the carbonation depth was between 5 mm and 30 mm.

YOKOYAMA et al. [26] carried out sustained exposure tests in which concrete elements were exposed to the climate in various ways. Some results of this research are shown in table 12.

Table 12. Carbonation depths at exposure tests.

| | carbonation depth (mm) | | | |
| | outside shielded | | outside, not shielded | |
	7 years	17 years	7 years	17 years
lightweight concrete with lightweight sand	5.1	6.5	13.0	14.2
lightweight concrete with normal sand	3.5	5.4	10.4	11.5
gravel concrete	2.9	4.3	8.5	9.5

It is obvious that the environment to which the test specimen were exposed must have played a decisive role. The type of concrete has much less influence. A direct comparison, however, is not possible because no detailed data on the composition of the mixture are available. Lightweight concrete with normal sand appears to be somewhat more unfavourable than gravel concrete, but practically the differences are of little importance. As is to be expected, lightweight concrete with lightweight sand is again somewhat more unfavourable than lightweight concrete with normal sand.

BERESFORD and HO [27] reported on laboratory research carried out on concrete with basalt, expanded soft slate and lava slag as coarse aggregates, with normal sand as fine aggregates. The test specimen were exposed to an atmosphere with an increased CO_2-content for a period of one month. The carbonation depths measured for the various types of concrete hardly appeared to differ. The depths varied from 10 mm to 18 mm, in which a clear relation to the water-cement ratio was established.

Table 13. Influence of the type of aggregate and of the amount of cement on the carbonation depth.

aggregate	cement content (kg/m³)	average carbonation depth (mm)
gravel	350	1.5
	204	13.0
sintered fly-ash	370	2.0
	230	13.0
expanded clay foam slags	380 - 440	3.0
	235 - 260	19.0

66

GRIMER [28] reported on an extensive series of exposure tests on various types of concrete. In this case the elements were exposed to a polluted atmosphere for a period of six years. The tests clearly demonstrated that the influence of the amount of cement is much more important than the type of aggregate. The most important data are summarized in table 13. It can be concluded that the cement content, the quality of the structure and the environmental conditions are the most influential factors with regard to the carbonation depths that are to be expected. With cement contents of about 350 kg/m^3 and more, carbonation will not be a problem with gravel concrete nor with lightweight concrete.

6.3 Inspections at existing structures

In the above-mentioned literature there are some reports on inspections carried out at existing structures. The ship Selma, for example, a 7500 tons tanker, built in 1919 of reinforced lightweight concrete with expanded clay as coarse aggregate, was sunk in 1922. The ship has been in seawater ever since and over a height of about one meter it is exposed to wet-dry cycles in salty air. In 1953 a number of drilled cores were taken from the hull. The cylinder compressive strength was between 55 N/mm^2 and 75 N/mm^2. The modulus of elasticity was 20000 N/mm^2. Despite the limited concrete cover on the steel (16 mm) there was hardly any corrosion. In 1980 a second investigation was carried out. The cylinder compressive strength was on average 70 N/mm^2 with a modulus of elasticity of 25000 N/mm^2. Still hardly any corrosion could be found.

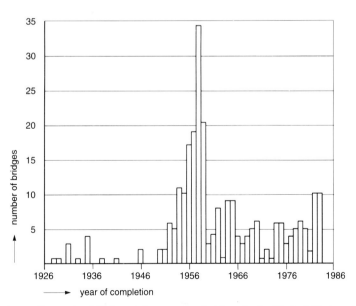

Fig. 38. Ages of examined lightweight concrete bridges in the United States of America.

Interesting data for lightweight concrete with much lower cube compressive strengths can be extracted from a report on the condition of 257 bridges of lightweight concrete published in 1985. The histogram in figure 38 gives a survey of the ages of these bridges.

On the basis of data acquired via inspection the following conclusions have been drawn:

- The analysis of the total number of inspected bridges shows that a good lightweight concrete guarantees a better durability than "any gravel concrete". Lightweight concrete seems to be less sensitive to the effect of thaw salts than gravel concrete.
- The irregularities as found in lightweight concrete are first of all due to the fact that the difference between lightweight concrete and gravel concrete was not really understood. For example, the relation between the water demand of the concrete mixture and the pore volume is not always properly recognized.
- So far the inspected lightweight concrete bridges have demonstrated a satisfying durability and the evidence increases that the durability is at least as good as that of gravel concrete. The few cases in which the behaviour was unsatisfactory are in all probability a result of poor detailing if not of poor control during execution.
- Lightweight concrete with a lightweight fine aggregate is less durable than lightweight concrete with natural sand. With the latter there is a minimum chance of continuous channel formation via contacts between the porous particles and optimal advantage is taken of the dense layer around the coarser aggregate particles.
- Although detailed descriptions of mixtures are not available, it is obvious that the amount of cement plays an important role with regard to the durability. Even if it should be a matter of "inferior" properties of the "lightweight aggregates" this will in any case be sufficiently compensated by the use of extra cement or some extra other very fine material.
- The favourable observations mentioned in the inspection report are made at structures of lightweight concrete with relatively low compressive strengths, which means that the issue of durability in the higher-quality types of lightweight concrete investigated by Research Committee C 75, will play an even less important role.

6.4   Conclusions

On the basis of the physical behaviour of lightweight concrete at corrosion there is no reason to assume that lightweight concrete would be less durable than gravel concrete. On the one hand it can be stated that the influence of porous aggregates on the corrosion process at the cracks will not differ significantly from gravel concrete. Lightweight concrete only requires some more corrosion product to halt the corrosion process. On the other hand, there is hardly any difference in the permeability of the concrete cover between lightweight concrete and gravel concrete. This is mainly caused by the fact that around the porous lightweight aggregate particles a highly homogenous and dense skin is formed. The conclusion that lightweight aggregate is not necessarily less durable than gravel concrete is supported by many research data and practical inspections. For further information on the literature study summarized above reference is made to [29].

CHAPTER 7

CALCULATION RULES FOR LIGHTWEIGHT CONCRETE

7.1 General

The new series of standards for building structures, the TGB-1990 series, includes the standard NEN 6720 "Technical Principles for Building Structures TGB 1990 - Regulations for Concrete. Structural requirements and calculation methods (VBC 1990)" for concrete structures. This standard contains a restrictive condition with regard to the aggregates, from which follows that the standard does not apply to structures of concrete with coarse lightweight aggregates. To fill in this gap in the regulations, supplementary regulations to NEN 6720 (VBC 1990) have been drawn up for concrete with coarse lightweight aggregates. These supplementary regulations have been included in CUR-Recommendation 39 "Concrete with coarse lightweight aggregates" [30] together with rules for design and calculation of structures of concrete with coarse lightweight aggregates.

The CUR-Recommendation only mentions those clauses from NEN 6720 that are supplemented, changed or entirely excluded. The contents of the CUR-Recommendation are based on literature research and on the research results of Research Committee C 75 [31]. They have been linked up as much as possible to the draft European standard for lightweight concrete structures [32]. The range of application of the CUR-Recommendation is restricted to concrete with coarse lightweight aggregates, consisting of particles with pelletised fly-ash conglomerate, blown clay particles or particles of sintered fly-ash, in as far as they meet the requirements for aggregates laid down in the recommendation. For concrete with coarse lightweight aggregates that deviates from these regulations, it will have to be established with the help of the procedure drawn up by CUR Research Committee C 79 "Concrete with gravel replacing aggregates", to what extent application of the recommendation – whether or not modified – is justified.

7.2 Contents of the CUR-Recommendation

The deviating behaviour of concrete with coarse lightweight aggregate has consequences for the material properties to be used, the calculation of the load distribution, the dimensioning and the verification, as well as for the detailing of the structure. Each of these subjects will be briefly discussed below.

Material properties
The deviations concern the value of the tensile strength, the modulus of elasticity (and consequently the stress-strain diagram) and the creep and shrinkage properties. The extent of the deviations is made dependent on the density of the hardened concrete in such a way that there will be no deviations if the density equals

that of gravel concrete (2300 kg/m³). The result is a sliding scale that links up with the properties to be used for gravel concrete.

By way of illustration, the changes in the mentioned properties are indicated below when the density is 1600 kg/m³:

- tensile strength   - 18 %;
- modulus of elasticity - 42 %;
- creep coefficient   - 42 % (for B 25 and higher);
- shrinkage   +20 % (for B 25 and higher).

Force distribution

The deviations involved in determining the force distribution concern the application of the quasi-linear-elasticity theory, the conditions for disregarding the second-order moments in braced frames and the determination of the total eccentricity for the dimensioning of columns. These deviations can be traced back to the lower modulus of elasticity of lightweight concrete as compared to gravel concrete, which leads to larger deformations. Here also the extent of deviation is made dependent on the density of the concrete, which means that a sliding scale is used, which is linked to the properties used for gravel concrete.

Dimensioning and verification

The dimensioning and verification rules in NEN 6720 (VBC 1990) have been adjusted on the subjects of shear force, punching, torsion, deflection and cracking. The changes with regard to shear force and torsion concern a reduction of the τ_2-value. As regards the shear force it is also laid down that the angle between the diagonal compression struts and the axis of the part of the structure concerned may not be assumed as small as is allowed for gravel concrete. As far as punching is concerned, it is not allowed to take into account the punching reinforcement, since the effect of punching reinforcement in lightweight concrete is insufficiently known. As regards the verification of the deflection, the use of the equivalent bending stiffness for the simplified methodis not allowed. Finally, the verification rules for cracking have been modified because of the different bond behaviour of lightweight concrete as compared to gravel concrete. Once again, all changes are related to the density of the lightweight concrete.

Detailing

The rules for detailing have been changed at various points because of the deviating properties of lightweight concrete. These changes concern the concrete cover, the detailing rules for reinforcing steel and prestressing steel, the application of transverse reinforcement and the magnitude of the bearing stresses. In a number of situations the concrete cover is to be increased with 5 mm. This concerns non-moulded surfaces, as far as these can be inspected, and structures in the environmental classes 3, 4 and 5. The changes concerning the detailing of concrete and prestressing steel and the application of transverse reinforcement result from the smaller splitting resistance of lightweight concrete. This leads to larger minimal radii for bends, whereas bundling is not allowed. In consequence of the different bond behaviour, larger anchoring, transfer and lap lengths have to be used. Also in relation to the smaller splitting resistance, the use of transverse reinforcement in structures pre-stressed with post-tensioned steel is obligatory. This transverse reinforcement is not required in "lightly" pre-stressed structures. Due to the smaller splitting resistance the magnitude of the bearing stresses is also limited. Finally, a number of detailing rules is not applicable to lightweight concrete. This concerns punching reinforcement, loop connections and cast-in anchors, since in these cases the knowledge on the behaviour of lightweight concrete is insufficient.

The smaller splitting resistance has previously been mentioned several times as a reason for changes. This smaller splitting resistance is caused by the lower modulus of elasticity in combination with the lower crack energy of lightweight concrete as compared to gravel concrete. Although the CUR-Recommendation includes a number of regulations with regard to this smaller splitting resistance, it cannot be excluded that these regulations are insufficient in some cases. Therefore the designer needs to be aware of the fact that supplementary measures may be required in situations in which the reliability of a structure or parts of a structure depends on splitting effects.

PRODUCT SPECIFICATIONS OF THE EXAMINED LIGHTWEIGHT AGGREGATES

Table A1. Product specification of Lytag according to data provided by the manufacturer.

Lytag	dimension	particle size fraction	
		4 - 8	6 - 12
dry density particles	kg/m^3	1420 ± 75	1400 ± 75
dry density loose material	kg/m^3	780 ± 65	760 ± 65
water absorption			
after 30 minutes	% ± kg/m^3	15 ± 40	15 ± 40
after 24 hours	% ± kg/m^3	18 ± 40	18 ± 40
particle strength	N/mm^2	> 5	> 4
particle distribution cumulative sieve residues			
C 16	% (m/m)	-	0
C 12.5	% (m/m)	0	≤ 15
C 8	% (m/m)	≤ 15	60 - 90
C 5.6	% (m/m)	-	> 85
C 4	% (m/m)	> 85	-
2 mm	% (m/m)	> 97	> 97
parts smaller than 63 μm	% (m/m)	< 2.0	< 2.0
amount of sulphate	% (m/m)	< 1.0	< 1.0
amount of chloride	% (m/m)	< 0.02	< 0.02
loss of glow	% (m/m)	< 5	< 5

Table A2. Product specification of Aardelite 1630 according to data provided by the manufacturer.

Aardelite 1630	dimension	particle size fraction 4 - 16
dry density particles	kg/m^3	1630 ± 75
dry density loose material	kg/m^3	1000 ± 65
water absorption		
after 30 minutes	% ± kg/m^3	11 ± 40
after 24 hours	% ± kg/m^3	18 ± 40
particle strength	N/mm^2	> 4
particle distribution cumulative sieve residues		
C 20	% (m/m)	0
C 16	% (m/m)	≤ 15
C 12.5	% (m/m)	30 - 60
C 8	% (m/m)	55 - 85
C 4	% (m/m)	> 85
2 mm	% (m/m)	> 97
parts smaller than 63 μm	% (m/m)	< 2.0
amount of sulphate	% (m/m)	< 1.0
amount of chloride	% (m/m)	< 0.02
loss of glow		not appl.

Table A3. Product specification of Liapor according to data provided by the manufacturer.

		particle size fraction 8 - 16 according to DIN 4226 part 2			
	dimension	Liapor 5	Liapor 6	Liapor 7	Liapor 8
dry density particles		-	-	-	-
dry density loose material	kg/m^3	500 ± 25	600 ± 25	700 ± 25	800 ± 25
water absorption					
after 30 minutes	$\% \pm$	7 ± 40	6 ± 40	7 ± 40	7 ± 40
after 24 hours	kg/m^3	-	-	-	-
particle strength	$\% \pm$	> 4	> 4	> 4	> 4
particle distribution sieve	kg/m^3				
residue on sieve	N/mm^2	0	0	0	0
25 mm	$\% \, (m/m)$	< 10	< 10	< 10	< 10
16 mm	$\% \, (m/m)$	> 85	> 85	> 85	> 85
8 mm	$\% \, (m/m)$	> 95	> 95	> 95	> 95
0.25 mm	$\% \, (m/m)$	-	-	-	-
parts smaller than 63 µm	$\% \, (m/m)$				
amount of sulphate	$\% \, (m/m)$	< 1.0	< 1.0	< 1.0	< 1.0
amount of chloride	$\% \, (m/m)$	< 0.02	< 0.02	< 0.02	< 0.02
loss of glow		-	-	-	-

REFERENCES

1. CUR-rapport 89-3, Lytag als toeslagmateriaal in beton. Evaluatie van toepassingen in Nederland. CUR, Gouda, 1989.
2. Experimenteel onderzoek naar het splijtgedrag van lichtbeton. TU Delft/TNO Bouw, TU Delft, rapport nr. 25.5-91-2, June 1991.
3. Numeriek vooronderzoek van op splijten belaste lichtbetonnen schijven. TU Delft/TNO Bouw, TNO Bouw, rapport nr. B-89-658, December 1989.
4. WURM, P. en F. DASCHNER, Teilflächenbelastung von Normalbeton Versuche am bewehrten Scheiben. Deutscher Ausschuss für Stahlbeton, Heft 344, Berlijn, 1983.
5. Numeriek onderzoek van op splijten belaste lichtbetonnen schijven. TU Delft/TNO Bouw, TNO Bouw, rapport nr. B-90-189, may 1990.
6. CORNELISSEN, H.A.W., D.A. HORDIJK en H.W. REINHARDT, Experiments and theory for the application of fracture mechanics to normal and lightweight concrete. In "Fracture toughness and fracture energy", Elsevier, 1986.
7. Parameteronderzoek van op splijten belaste lichtbetonnen schijven. TU Delft/TNO Bouw, TNO Bouw, rapport nr. B-90-770, April 1991.
8. Experimenteel onderzoek naar het splijtgedrag van excentrisch belaste ongewapende prisma's. TU Delft/TNO Bouw, TU Delft, rapport nr. 25.5-92-4, Juy 1992.
9. Experimenteel onderzoek naar het aanhechtgedrag van wapeningsstaven in lichtbeton. TU Delft/TNO Bouw, TU Delft, rapport nr. 25.5-91-3, July 1991.
10. Deformation-controlled uniaxial tensile tests on concrete. TU Delft, report nr. 25.5-89-15, 1989.
11. Numerieke simulatie van op trek belaste staven. TU Delft/TNO Bouw, TNO Bouw, rapport nr. B-90-769, April 1991.
12. Experimenteel onderzoek naar de sterkte van overlappingslassen in lichtbeton. TU Delft/TNO Bouw, TU Delft, rapport nr. 25.5-92-5, July 1992.
13. Experimenteel onderzoek aan balken met afschuifwapening (2 delen). TU Delft/TNO Bouw, TU Delft, rapport nr. 25.5-91-6, May 1992.
14. Numerieke simulatie van op afschuiving belaste lichtbetonnen I-liggers. TU Delft/TNO Bouw, TNO Bouw, rapport nr. B-91-0660, November 1991.
15. Parameteronderzoek van op afschuiving belaste lichtbetonnen I-liggers. TU Delft/TNO Bouw, TNO Bouw, rapport nr. B-92-0243, June 1992.
16. CASTENMILLER, G.J.J. en A.J.M. SIEMES, Invloed van lichte toeslagmaterialen op de corrosie van de wapening in beton: een modelmatige benadering. TNO-rapport nr. B-91-0287, April 1991.
17. HOLM, T.A., T.W. BREMNER en J.B. NEWMAN, Lightweight aggregate concrete subject to severe weathering. Concrete International, June 1984.
18. KHOKHRIN, N.K., The durability of lightweight concrete structural members. Kuibyshev, USSR, 1973.
19. PODVALNYI, A.M., Phenomonological aspects of concrete durability theory. Materials and Structures, Research and Testing (RILEM, Parijs), nr. 51, May-June 1976.

20. Nishi, S., A. Oshio, T. Sone en S. Shirokuni, Watertightness of concrete against sea water. Journal, Central Research Laboratory, Onada Cement Co., Tokio, nr. 104, 1980.

21. Haynes, H.H., Permeability of concrete in sea water. Performance of Concrete in Marine Environment, SP-65, American Concrete Institute, Detroit, 1980.

22. Lydon, F.D., Working Group on Lightweigth Concrete, Final Report. Proceedings of the Tenth Congress of the Fédération Internationale de la Précontrainte, Volume 3 (Technical Paper), New Delhi, 16 - 20 February 1986.

23. Guide to the structural use of lightweight aggregate concrete. The Institution of Structural Engineers, The Concrete Society, October 1987.

24. Walsh, R.W., Restoring salt-damaged highway bridges. Civil Engineering - ASCE, nr. 5, May 1967.

25. Ohuchi, T., M. Hara et al., Some long-term observation results of artificial lightweight aggregate concrete for structural use in Japan. RILEM/ACI International Symposium on Long-Term Observation of Concrete Structures, 17 - 20 September 1984, Budapest.

26. Yokoyama, M., Y. Kitamura en T. Yamashita, Artificial lightweight aggregate concrete exposure tests, Part 5. Concrete quality in seventeenth year of exposure. Transactions of the Architectural Institute of Japan, Summary of Technical Papers.

27. Beresford, F.D. en D.W.S. Ho, The repair of concrete structures – a scientific assessment. Concrete Institute of Australian Biennial Conference 1979, Canberra.

28. Grimer, F.J., Durability of steel embedded in lightweight concrete. Concrete, nr. 4, april 1967.

29. Walraven, J.C., Duurzaamheid lichtbeton: een literatuurstudie. TU Delft, december 1993, rapport nr. 25.5-93-7, april 1994.

30. CUR-Aanbeveling 39, Beton met grove lichte toeslagmaterialen. Aanvullende bepalingen op NEN 6720 (VBC), NEN 5950 (VBT) en NEN 6722 (VBU), juli 1994.

31. Achtergronden bij de constructieve aspecten van de CUR-Aanbeveling "Beton met grove lichte toeslagmaterialen". Adviesbureau ir. J.G. Hageman B.V., rapport nr. 2498-1-3, oktober 1993.

32. prENV 1992-1-4, Eurocode 2: Design of concrete structures, Part 1-4: The use of lightweight aggregate concrete with closed structure. CEN, februari 1993.